Build Your Own Solar Heating System

KENNETH CLIVE

Lucerna Publishing, Minneapolis, Minnesota

Build Your Own Solar Heating System

by Kenneth Clive

Published By:

Lucerna Publishing
2936 34th Avenue South
Minneapolis, MN 55406-1708 U.S.A.

Copyright © 2007 by Kenneth Clive and Lucerna Publishing
1st Edition
ISBN, 978-0-9754236-2-2

Library of Congress Control Number: 2007901061

Printed in the United States of America

Acknowledgements

There are many people to thank for the support I have received in the writing of this book. This book would not have been possible if I had not had the first-hand experience of building my own solar heating system. Whenever I decide to embark on a new project I have the support of my immediate family. I am lucky to have neighbors willing to lend a hand when I am unable to perform a task single-handedly. On a particularly windy day my ladder blew down and I was stuck on the roof with no way to get down until my next door neighbor heard me yelling and came by to put the ladder up again.

My neighbor across the street from me assisted with some much needed muscle power to drop the steel enclosure needed for the reservoir in place. I also received his assistance while hauling the collector panel onto the roof.

Thanks to my daughter for all the help during the long and tedious process of digging the pit for the reservoir. Between us, we moved about 5 tons of dirt. My sister, who works at the New York Board of Education and also teaches computer courses at Community Colleges, took time out of her busy schedule to read and offer many valuable suggestions.

Finally, I thank my wonderful wife for putting up with the countless days of the construction of the project and the time needed to produce this book because of which I was often unavailable to help with household chores.

iv

Contents

About the Author

Kenneth Clive has lived in Minneapolis, Minnesota for the last twenty six years where he has spent a lot of time practicing his love for designing and building many different kinds of projects. Kenneth has a Masters degree in Physics with an emphasis in electronics. He grew up in the port city of Karachi in Pakistan and after acquiring his degree spent seven years as the Assistant Director of the YMCA Technical Institute where, in the mid 1970's he first learned about a simple solar thermosiphon water heating system demonstrated by a representative of one of the international sponsors of the institute. He immediately saw the possibilities of using the sun's energy for solar desalination of water for drinking and heating.

At the time an interest in alternative energy was growing in many countries around the world and designs for solar cookers using parabolic reflectors were showing up. The United Nations was involved in helping villages in countries around the world where electricity and gas were unavailable or unaffordable, to promote alternative energy techniques to build small biogas units and solar cookers.

Ken has taught Math, Physics and Computer courses at the college level both in Karachi as well as in Minnesota. Although he has worked as a computer consultant for many years with several large corporations including 3M, Honeywell, US Bank, Target, and Guidant, he has a passion for self-sufficiency and eco-friendly power production.

Ken put his technical knowledge and deep interest in alternative energy projects to build the solar heating system described in this book. The idea behind the project was to build it with easily available tools and resources so that it could be duplicated by anyone who is handy with basic tools and construction techniques.

Preface

This is an exciting time to be involved in solar technology. There is a renewed and growing interest in applying techniques which have been around for a long time but have been ignored due to the cheap and easy access to fossil fuels. Nuclear power plant accidents and the suppression of the truth about its dangers have made people wary of giving permission to build new ones. Coal fired plants along with our love of the automobile and its dependence on petroleum to run it have created the ecological disaster we see today.

Each of us shares a deep responsibility for the legacy we leave behind for future generations. There have been so many wonderful technological advances which can be applied to making use of the sun's energy to supply our needs and make life comfortable. Many still think that if they make the move to solar energy they will have to endure tremendous discomfort. Once they see how "normal" a solar house can be the interest level begins to rise. Lots of people have seen former vice president Al Gore's movie "An Inconvenient Truth" about global warming but don't really know what to do to begin the changes in lifestyle needed. Tons of resources are now available to get educated about what's happening and what you can do.

Switching over to a more ecologically sound lifestyle should be guilt free. However, it is sort of like going on a diet. The right diet is usually not painful but you do have to start. The changes won't happen by themselves. Find out all of the ways your home uses and loses energy. What are your requirements? Finally, what can you do to reduce your energy consumption. Your energy provider can, most likely, provide you with an energy audit to show you where you may be wasting it. So go ahead, take the plunge. Learn more about what you can do. You'll be glad you did.

Warning - Disclaimer

Chapter 1

Why Go Solar?

1 | Why Go Solar?

If there's one thing that has been self evident since the beginning of this millennium, it is that we cannot depend on utility companies to provide us with stable prices for our energy needs. It isn't entirely their fault. In a market based economy, businesses will sell and charge what people are willing to buy and pay. An interest in alternative forms of energy is nothing new. In the United States people have become accustomed to paying some of the lowest prices for electricity, natural gas and gasoline, anywhere in the industrialized world. It takes man-made or natural disasters every few years to shake us out of our complacency and realize that we should give serious consideration to energy that is less dependent on fossil fuels. In 2005 hurricane Katrina destroyed major petroleum processing plants in Louisiana and Mississippi and the oil industry immediately took advantage by sending prices of gasoline soaring and making tens of billions of dollars in profits. Depending on the kind of relationship that the United States has had with petroleum producing countries particularly in the Middle East has determined the price that consumers have paid for everything from vacation travel to food items. Petroleum is inextricably intertwined in every part of life in an industrialized society, from the asphalt in the roads we drive on to the packaging of the food we eat. In the northern climates a huge amount of energy is used simply to heat our homes and water. Using solar energy as an alternative to oil and gas can substantially reduce the amount you pay to stay warm.

We all know that being in the sun makes us feel warm and good. There are ways to harness the energy that gives us this warmth reliably. Solar energy is free and is our only truly renewable resource. You essentially have free energy for heating, cooling and electricity for as long as the sun rises in the east and sets in the west. There is an initial cost of capturing and converting this energy into a useable form. There is also a nominal cost of maintaining the equipment. However, aside from that, once your system is up and running the cost is recovered over a period of time.

All of us know that the sun's heat can be intense enough to cause highways to buckle sometimes. People often choose light colored vehicles in hot climates because darker colors tend to get much hotter. On extremely hot days we are warned not to leave children or pets in parked cars in the sun. The temperatures quickly rise to dangerous levels. Knowledge of these basics allows us to come up with ways to capture this fabulous heat source and make it work for us.

The principle behind capturing and using solar energy is actually quite simple. Devices known as solar collectors capture the heat from the solar radiation. The heat collected is stored in a heat reservoir until it is needed. This heat is then transferred to the item to be heated either passively, using natural principles of heat transference, or actively by means of pumps or fans.

If you are relatively handy with some basic construction techniques you can build your own solar heating system which can include heating your house in winter and providing hot water. Of course you can also capture solar energy to convert to electricity to run many of the appliances you have in your home. This book is dedicated to showing you how to capture and use solar energy in the form of heat. After many years of toying with the idea, I finally decided to go ahead and build my own system using the information I acquired. I hope you will be encouraged to build your own solar heater after reading this book. Being an avid do-it-yourselfer, I am not afraid to try new things. Building such a system, however, requires a huge amount of labor since there are many stages to completing the project. It took me more than four months working almost full time with help from my daughter during the digging of the reservoir pit. For those of you who may not be as confident in your own abilities I have included resources you can use to complete your project.

It is encouraging to see that more people have become aware of the harmful effects of carbon emissions produced by industrialized countries and that they are starting to realize that something has to be done. The task seems overwhelming but each of us doing our own small bit can make a huge impact. We're all in this together.

Chapter 2

How it Works

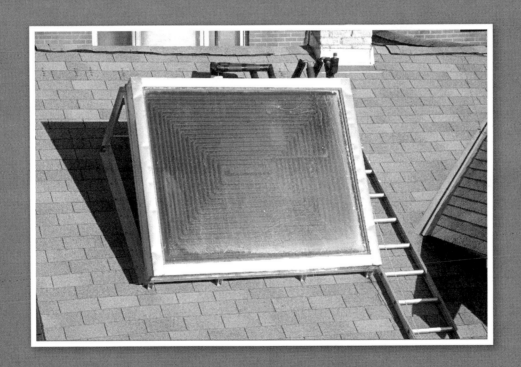

2 | How it Works

In this chapter I will give you a general idea of how any solar heating system works and then tell you a bit about the one I constructed. I knew the basics from my background in Physics and had a general concept of what I wanted to do. The hard part was in the application of the concepts to the implementation in physical form. I often found that I needed to describe what I wanted to accomplish to a supplier of parts and materials and managed to find the right items that I needed. Even though this is a general description, you should read it all the way. It may spark ideas of your own. If you feel you would like to do something differently from the way I have described, you should obtain any technical data regarding the changes you are planning. Each individual section is not complicated but the system as a whole needs to function properly with all the sections combined. I have stressed the importance of adequate safety measures several times throughout this book. Do not take them lightly. If you are an avid do-it-yourselfer then you may be familiar with a program on Public Television called "The New Yankee Workshop" hosted by Norm Abrams. One of the things he always says before beginning any project on that show is, "Be sure to read and understand all the safety rules that come with your power tools..." This is a very good piece of advice to follow, since it is applicable to more than just power tools.

2.1 Many Different Ways To Do It

There are many different ways to build a solar heating system. This book shows you how I built mine and how you can build a similar system. I would encourage you to do your own research, both on the internet as well as in your local library. There is quite a bit of material available but most of it does not go into details of how and where to obtain the components needed. I have attempted to explain terms which you may not have come across in your experience, however there may be terms I did not explain since I considered them part of a do-it-yourselfer's general vocabulary. I have read too many books and articles online that make passing references to terms without explaining what any of them mean in the context they are being used. They also often fail to mention details regarding quantitative or qualitative measurements. As hard as I have tried to be comprehensive in my descriptions, I am sure that someone will find something that I did not cover adequately. Constructive comments are welcomed at the address of the publishing company.

Solar heating systems have been around for a long time. They all depend on some means of capturing the sun's heat and storing it for use at a later time. Heat always travels from a warmer area to a cooler area. It does this in three different ways, radiation, conduction and convection. Radiation is the movement of heat waves through space from a warmer object to a cooler one. Heat waves are electromagnetic energy. In conduction, heat travels through a substance by causing a vibration in the molecules of the substance through which it is travelling, and in convection heat travels due to the circulation of heated air. Warm air rises. As it rises it loses its heat to objects surrounding it. When it cools it becomes dense and sinks. Passive systems use this principle. Active systems use mechanical means such as a fan or pump to send the heat where it is needed.

Different solar heating systems may employ one or more of these principles to achieve the desired end result. The heat collected can be used for a variety of purposes including heating your home in colder climates and heating water for a water heater so that the water heater uses less energy because it does not have to turn on as often.

2.2 How My System Operates

I designed my system around some basic principles of heat collection and transference. A solar panel that I am referring to as a collector, collects heat from the sun on sunny days all year round and transfers this heat to an underground insulated heat reservoir filled with rocks. The rocks absorb the transferred heat and retain it until it is drawn off to pre-heat water for the water heater and to heat the house in the cold weather. Each of the elements of the system is described next to give you a better understanding of how my system operates. Throughout the rest of the book you will find detailed instructions on how to build each of these elements and combine them into a fully functional heating system.

2.3 The Collector

The collector, as its name suggests, collects the suns energy on sunny days. This is done by laying a coil of copper pipe in an insulated box and painting both the inside of the box and the pipe coil with a flat black paint. A principle of Physics published in 1901 by the physicist Max Planck, tells us that black objects absorb more heat than lighter colored objects. We don't need to know all the mathematics behind his publication in order to take advantage of it. You may be interested to know that Planck presented his publication five years before Einstein. This is just a piece of trivia you can file away in case you ever happen to be a contestant on a TV quiz show. Also, flat black absorbs more than glossy black since it does not reflect light or heat. The coil of copper pipe is filled with non-toxic antifreeze (propylene glycol) and pumped through the system using a circulation pump.

The copper pipe heats the antifreeze flowing through it much the same way as antifreeze flowing through the engine block in a car. In a car the heat absorbed by the antifreeze is circulated through the radiator and cooled by the radiator fan in order to keep the engine cool. In our heating system the heat collected is sent using the antifreeze as the carrier down to the reservoir where it is stored for use later. The collector is mounted on the roof of the house, preferably where it is least subject to shadows from surrounding buildings or trees.

The pipes exit the collector and are routed through the roof and interior walls of the house down through the basement and eventually into an underground pit outside the basement wall that houses a reservoir. The antifreeze in the pipes is pumped through while the sun is out and heating the collector. Electronic circuitry is used to control when the pump is turned on and off.

2.4 The Reservoir

Collecting all that heat doesn't do us any good if we don't have a way to store it and use it when we need it most. The word reservoir is most often used in the context of a system for water storage. This is easy to visualize. Heat, on the other hand, is electromagnetic energy. You can feel its effect but it isn't something you can hold in your hands. However, because heat can be transferred between objects by convection, conduction, or radiation, we can store it by transferring it to some material with good absorption and retention properties inside a well insulated space.

All insulating material has its own rate of heat loss. The resistance to heat loss of a material is referred to as its **R** value. If you've ever purchased insulation for your home, you've heard of this. The reservoir I built was below ground. I did this mainly for aesthetic purposes. I did not want to have a huge object sitting next to my house above ground. Any option you choose will have its own set of advantages and disadvantages. Since I chose to have my reservoir below ground, I had to use insulating material that would not be ruined, or lose its insulating properties by moisture. I chose foam board insulation because it is so easy to work with.

The material I chose to retain the heat collected was 2 inch rock which is available at landscape supply places. I used my own truck to haul the rock because I did not want several tons of rock sitting in my backyard waiting for me to get to it. I happen to own a Ford F-150 pickup truck with some heavy duty springs and tires. My system required about four tons of rock so I accomplished this by making four trips to the landscaping company and getting about a ton at a time. At the yard they let you load your truck yourself or for a very nominal fee they have a big front-end loader drop a scoop into your truck bed. It takes forever trying to shovel in a spadeful at a time, not to mention the physical effort. I just let them load my truck for me. You can also have the rock delivered to your house by the company if you pay an extra amount for the delivery. Rock is a great transfer material. You will never need to worry about leaks, as you might in as system that uses water as the transfer medium.

2.5 The Exchanger

The rock in the reservoir used to store heat also functions as the heat exchanger. In order to avoid contamination of the water supply an isolated coil of pipe is used to draw the stored heat to pre-heat the water for the water heater. The water from the city supply is first routed through the pipe coil in the reservoir where it is warmed by the surrounding rock before going to the inlet of the water heater. Another isolated coil filled with non-toxic antifreeze is used to draw heat to warm the house. Keeping the coils isolated from each other is the best way to prevent the fluids from coming in contact with one another.

Once your system is up and running you will find that the temperature in the reservoir varies on a daily basis. The changes are from the length of time that the sun is able to heat the collector and the amount of heat being drawn off by the water heater and house heater. If you happen to have a stretch of several days where the sky is overcast, and the reservoir cools off, it will take time to regain the stored heat. In the next chapters we will get into the details of the construction of our heating system. There is a considerable amount of work involved in putting it all together, but if you are patient the results are worth the effort.

Chapter 3

Building the Solar Collector

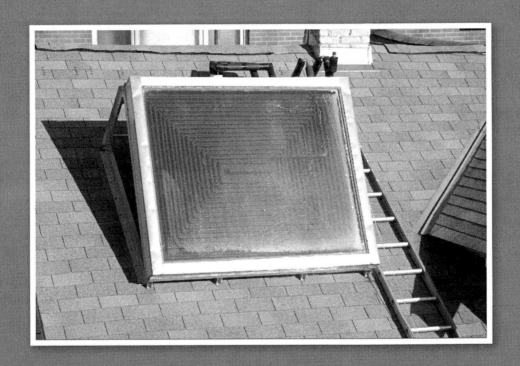

3 | Building the Solar Collector

The solar collector is, in effect, a small greenhouse. Sunlight consists of small electromagnetic waves which can easily pass through transparent materials such as glass. Opaque objects that are black absorb these waves and become hot. The heat waves being longer than light waves become trapped and are unable to escape back out through the glass. This is the principle behind the solar collector. The solar collector described in this book is a flat insulated box covered with two sheets of glass. The glass sheets are separated by a spacer similar to double glazed windows. Double-walled acrylic panel is much lighter than glass and easier to manage. Double-walled acrylic panel is a bit difficult to find and fairly expensive compared to glass. I used glass because I used to have a glass shop next door and the owner was discarding a large double-glazed window that had some minor flaws. However, I highly recommend the use of double-walled acrylic. I painted the inside of the box a flat black. The absorber consisted of 3/4 inch, Type L, copper pipe which was also painted a flat black after assembly. Don't use glossy paint. Flat black paint absorbs heat much better than glossy paint.

The entire collector along with the copper pipe and glass cover can be quite heavy depending on its size. I recommend building a few smaller collectors connected together rather than one large one. The reason for this is that after construction the collector has to be hoisted up onto your roof and that can be tricky and possibly dangerous if the collector is heavy and you do not have one or more people to help you.

3.1 Constructing the Outside Box

How many collector panels do you need? After reading this book you can decide for yourself. The photos shown are for reference only. The dimensions given below are the ones you should use. Since every house has different heating needs you can add additional collectors as necessary.

The solar panel without the glass installed. The handles were attached temporarily to help moving it around before being permanently installed.

Materials you will need:
1" by 6" pine boards
2" by 4" lumber
¼" plywood
2" Pink high-density styrofoam insulation
1" galvanized all-purpose screws
1½" galvanized all-purpose screws
3" galvanized all-purpose screws
1" brads
Construction adhesive such as Liquid Nails

PL300 foam board adhesive
¾" Type L copper pipe
¾" 90 degree copper elbows
¾" copper saddle clamps
Silver solder
Lead-Free Tinning flux

Tools you will need:
Power circular saw
Table saw
Miter box
Finish quality saw blade
Power drill
Set of drill bits
Screwdriver bits
1 inch hole saw
Measuring tape
Sharpie fine-tip black permanent marker
Sharp pointed pencil
Copper pipe cutter
Propane torch
Carpenter's square
Hand saw
Caulking gun

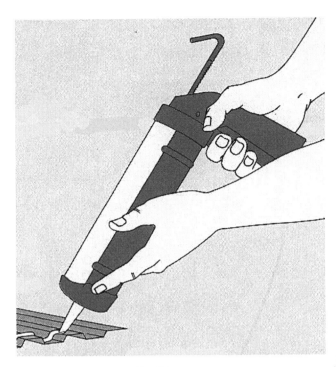

Caulking gun

Construct the outside box using 1 inch by 6 inch lumber and ¼ inch plywood. The external dimensions are 30 inches by 48 inches. 1 by 6 lumber when purchased at a lumber yard is actually ¾ inch by 5 ½ inches. With the ¼ inch plywood attached the height of the box will be roughly 5 ¾ inches. The construction adhesive comes in tubes which should be the same size as the caulking gun. A caulking gun is indispensable when applying adhesive over large areas.

Cut two lengths of 1 by 6 board 48 inches long. Cut another two lengths of 1 by 6 board 28 ½ inches long. Apply a small amount of construction adhesive to the ends of the 28 ½ inch boards. Butt the 48 inch long pieces against the ends of the 28 ½ inch pieces and screw together using the 1½ inch all-purpose screws. Check the box for square with a carpenter's square. Double check by measuring the box diagonally. Both diagonals should be equal. If you measured correctly, the 1 by 6 frame should be 48 inches long and 30 inches wide.

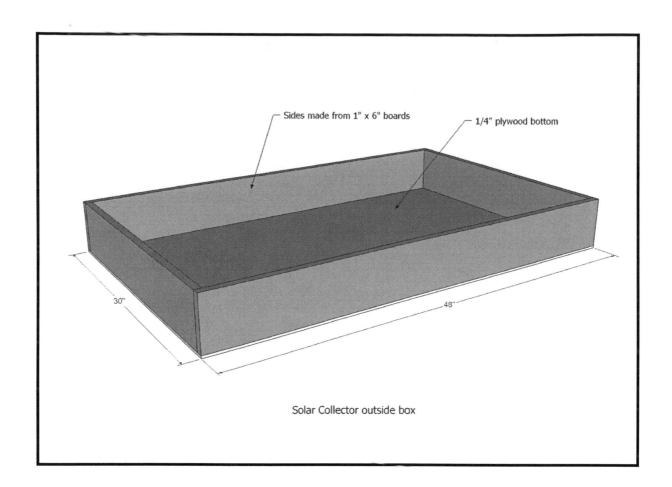

Sides made from 1" x 6" boards

1/4" plywood bottom

30"

48"

Solar Collector outside box

Cut a piece of ¼ inch plywood 30 inches wide and 48 inches long. If you bought a full 4 foot by 8 foot sheet of plywood you won't need to measure the 48 inches. Just measure 30 inches along the length of the plywood and cut it using your circular saw. A good quality saw blade will ensure a cleaner cut. Apply adhesive to the edges of the box frame. Lay the plywood on the frame making sure that the edges meet. Screw the plywood down using the 1 inch all-purpose screws. The screws should be approximately 6 inches apart.

3.2 Adding Insulation

Styrofoam insulation comes as white or pink and in a few different thicknesses. The pink insulation is denser and cuts cleaner than the white variety. Purchase the insulation in 4 foot by 8 foot sheets. For this part of your project one sheet of 2 inch thick foam will be sufficient. You can cut the insulation with a hand saw. Alternatively, you can carefully measure and mark both sides with a fine tipped marker then score both sides with a utility knife. The foam easily breaks along the score lines. Use whichever method you prefer.

Cut one piece of insulation measuring 46½ inches by 28½ inches. Cut two pieces measuring 46½ inches by 3½ inches. Cut two more pieces measuring 24½ inches by 3½ inches. Apply PL300 foam adhesive to the inside of the box constructed in the last section. Lay the 46½ inch by 28½ inch piece of foam at the bottom of the box. Apply foam adhesive around the perimeter of this foam piece and insert the remaining foam pieces. Press down firmly all around so that the foam bonds with the wood frame as well as the other foam pieces.

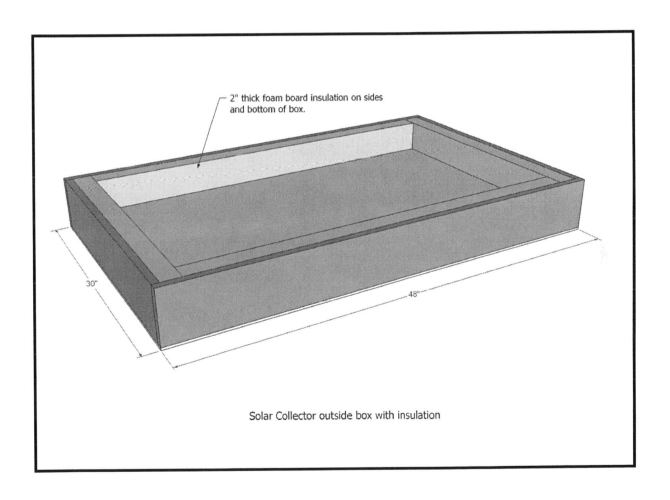

2" thick foam board insulation on sides and bottom of box.

30"

48"

Solar Collector outside box with insulation

3.3 Constructing the Inside Box

A box that fits inside the foam insulation has to be constructed next. This is the box that will hold the absorber made with copper pipes. The walls are made of 2 by 4 lumber and the bottom is made of ¼ inch plywood.

Cut two 42½ inch lengths of 2 by 4 lumber and another two 28½ inch lengths. Using a table saw cut a dado in each of the pieces ¾ inches wide and ¾ inches deep. This is easily done by some careful setting of the table saw. Set the fence ¾ inches from the saw blade and set the blade height to ¾ inches. Two passes on each piece will give you the correct dadoes. Use a miter box saw to miter both ends of all four pieces at a 45 degree angle. Join the pieces to make a frame as shown using construction adhesive and 3 inch all-purpose galvanized screws.

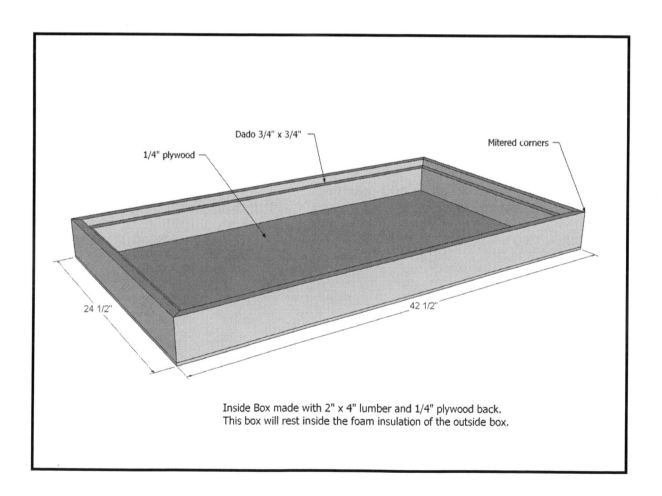

Inside Box made with 2" x 4" lumber and 1/4" plywood back.
This box will rest inside the foam insulation of the outside box.

Cut a piece of ¼ inch plywood measuring 42½ inches by 28½ inches. Attach it to the frame on the opposite side from the dadoes using construction adhesive and 1 inch galvanized all-purpose screws. Apply PL300 foam board adhesive to the inside of the foam box constructed in the last step and insert the inside box just constructed applying firm pressure so that good adhesion is obtained. When finished the entire assembly should be similar to the sketch shown on the previous page.

Cut four strips of ¼ inch plywood 3½ inches wide. Cut two of these strips 48 inches long and two 30 inches long. Apply construction adhesive to the top edges of the outside and inside box and apply the plywood strips securing them with 1 inch brads.

After the adhesive has had time to set, paint the inside of the box with flat black paint. Paint the outside any color you like. Paint is necessary to protect the wooden box from weather. Remember that it is going to be outside for many years in all kinds of weather. I took the extra precaution to clad the top and sides of the box with aluminum flashing. Aluminum flashing comes in rolls of varying width. I purchased a roll that was 1 foot wide and long enough to cover the entire perimeter of the panel. I used ¾" wire

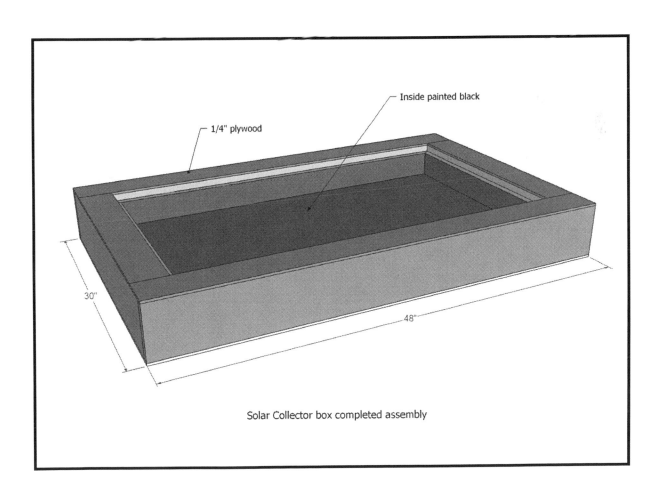

Solar Collector box completed assembly

nails to hold the flashing in place. Apply a bead of silicone caulk under the edges of the aluminum before nailing it down to prevent any moisture from creeping underneath.

3.4 Assembling the Absorber Pipes

We are referring to this part of the assembly as the absorber pipes. These pipes will absorb the solar energy that enters our solar collector box in the form of heat. Copper pipes that are part of the plumbing in most homes use type M copper pipes. We will use type L copper pipes throughout our project. Type L pipes have a thicker wall than type M. Type L copper pipe costs more than type M but is worth the extra cost.

This part of the project involves making many solder joints. Soldering copper pipe is not difficult but does require a bit of practice to ensure that your joints will not leak.

CAUTION: DO NOT attempt this unless you are sure of your technique. If you do not have absolutely perfect copper joins throughout, you run the risk of leaking. Antifreeze in the form of Propylene Glycol, is used in this project. The kind of antifreeze used in cars is highly toxic due to additives used to prevent corrosion of the engine block. **Do not use automotive antifreeze**. Propylene Glycol is actually non-toxic used in products like hand lotion. We will use non-toxic Propylene Glycol available from plumbing supply stores. Caution is also necessary when using a propane torch. Carelessness could cause you to burn down your house. Once again, if you are unsure of your abilities, find someone who can do it for you.

3.4.1 How to Use a Propane Torch

Propane torch kits are available at most hardware stores. One such kit includes a small propane tank, a combination valve and nozzle, and a striker. The kit may also include flux and solder. Check the flux and solder to make sure it is silver solder and lead-free tinning flux. If not, get it separately. You'll find it in the same area of the hardware store.

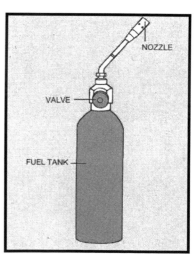

Propane Torch

Assemble the torch according to the instructions in the kit. Open the valve slightly and use the striker to ignite the propane. Opening the valve too much will blow the flame away. When the torch lights, open the valve until the flame is large enough to heat the area you will work on.

Hold the torch as close to upright as possible while you are working. If the tank is tilted too much the liquid propane will flow into the valve, blocking it so the flame goes out. If this is not possible you can get a kit that has a rubber hose extension that fits on the propane tank. This allows you to set the propane tank down so it is upright while you are free to move the hose for a more convenient angle.

If the area to be heated is close to wood or other flammable material cover the area with a flame retarding shield, also available at the hardware store. Make a habit of shutting off the torch every time you set it down. The flame is nearly invisible and can cause severe damage if placed near something that can catch fire.

3.4.2 Sweating a Copper Joint

The process of soldering a copper fitting to a copper pipe is called sweating. The process is simple but requires a bit of practice to get it right every time. Scour the end of the copper pipe with fine sandpaper or emery cloth. Next scour the inner surface of the fitting to be attached to the pipe with a wire brush. Once the two surfaces are clean, do not touch them with your fingers. Oil from your fingerprint will weaken the join. Apply a thin coating of flux to both surfaces with a small utility brush and slide the fitting onto the end of the copper pipe.

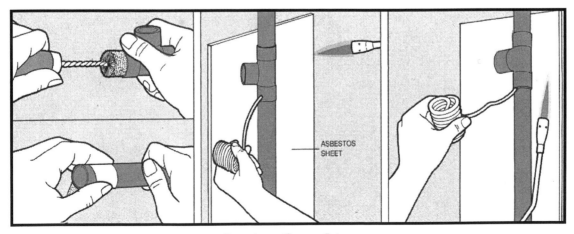

Sweating a Copper Joint

Heat the area with the propane torch until it is hot enough to melt the solder. Do not use the torch to melt the solder. When the join being soldered is hot enough, solder will melt on contact and flow evenly. If the solder does not melt, remove it and heat the area some more. When properly heated the flux will draw the molten solder into the fitting and seal the join. Continue feeding the solder until a bead appears around the rim and just starts to drip. Allow the join to cool before handling the pipe.

Let's get back to assembling the absorber pipes. Cut one each of the following lengths of ¾ inch copper pipe using a pipe cutter.

Group 1: 24", 22", 20", 18", 16", 14", 12", 10", 8", 6", 4", 2".
Group 2: 38", 36", 34", 32", 30", 28", 26", 24", 22", 20", 18", 16", 11".

Start with the inside loop using the 16" and 18" pipes from group 2 and the 2" and 4" pipes from group 1. Join the 16" pipe to the 2" pipe with a ¾" elbow, then join the 18" pipe to the other end of the 2" pipe with a ¾" elbow. Join the 4" pipe to the other end of the 18" pipe. Continue joining the group 2 and group 1 pipes with copper elbows. The last pipe joined should be the 24" pipe. Solder a ¾" threaded male adapter to the 24" pipe.

Solder one street elbow and one regular elbow to the innermost 16" pipe so that you are able to connect the remaining 11" pipe parallel to the outermost 24" pipe. Solder a ¾" threaded male adapter to the 11" pipe.

Copper elbow, tee, female pipe thread adapter
Saddle clamp, union, male pipe thread adapter

Place the entire assembly inside the collector panel box and using a 1½" hole saw cut holes in the box to match the positions of the 2 pipe ends. This is where the pipes will exit the box.

Cover the threads of the male adapters with masking tape and paint the entire assembly with flat black paint. When the paint is dry, attach the assembly to the box using the ¾" saddle clamps. The picture below shows how my assembly turned out. Yours should look similar to this except for the dimensions.

3.5 Attaching the External Pipes

We now need to connect a short length of pipe to each of the ends that should be close to the holes you cut in the box. Cut 2 lengths of pipe long enough so that they protrude outside the box at least 9" when inserted in the hole to meet the ends with the male pipe connectors. Solder ¾" female pipe adapters to one end of each of these two pipes and a ¾" union connector to the other end. Make sure that it is the end with the male thread. It will make it easier later on when you join this assembly to the rest of the system. Apply a thin layer of Teflon pipe thread compound to the threads of the female adapters. Using a pipe wrench or a pair of channel-lock pliers thread the pipes onto the ends of the assembly in the box and tighten them.

This entire assembly will be set aside until we are ready to install it on the roof. Use insulating spray foam to fill the gap between the hole in the box and the pipe protruding through. Cover the entire box with plastic tarp to protect the assembly and set it aside somewhere out of the way.

Chapter 4

Building the Heat Reservoir

4 | Building the Heat Reservoir

You have surely figured out by now that there is considerable manual labor involved in the construction of your solar heating system. You will have to find some way to keep yourself entertained during the next phase of construction. It involves digging the pit and unless you have several friends helping you it is going to take a while. I dug mine with the help of my daughter and we listened to story book tapes and CDs to make the job more fun. I have inserted pictures showing the progress of the digging process during my project to help you get an idea of what's involved.

4.1 First You Dig

Choose a convenient spot next to your basement wall. Your house does not have to have a basement. The spot you choose should be such that you have clear access to it from inside the house where it will not interfere with anything else. The reason for this is that several pipes will be coming through the wall later and the pumps, thermometers and controllers will be mounted against the interior of the basement or any other wall you choose. Another consideration is the amount of space on the outside of the basement wall. You need to have a 6 foot by 6 foot area right against the wall to dig. Make sure that it is not located where you need to have people traffic. If you anticipate that it could be a hazard, protect the perimeter with highly visible plastic fencing or other means so that no one will fall in by accident. You will be digging a hole 6 feet deep and the chance of someone falling in becomes highly likely unless you mark the area well. **Read the rest of this chapter before you proceed.**

4.1.1 Before You Dig

Before you dig, you should also call your local utility companies and have them mark the area for underground cables, and water and gas pipes. In my city there is only one number to call for marking both water and gas lines. The electric lines are above ground so I did not have to worry about them. In some cases electric lines may be buried underground so you must make sure you know where they are. The utility companies usually provide this service free of charge because of the great danger that you may accidentally hit a buried electric cable or a gas line. **This is extremely important. Do not take short cuts.**

Find a convenient location for piling up or otherwise distributing the dirt that will come out of the pit that you dig. You'll be surprised at the amount of space the loose soil takes up.

4.1.2 Lots of Digging

Once you have determined where the buried cables and gas lines are, you can decide where you plan to have your heat reservoir. Unless you have people helping you, this part is the most time consuming. Digging with a spade will take longer but is better than using machinery because you will be able to have straighter sides to your

pit without disturbing the earth around your site. Your goal is to dig a pit that measures 6 feet x 6 feet square horizontally and 6 feet deep vertically.

In order to keep the sides and edges from crumbling as you step in and out of the area, construct a frame with 2" x 8" lumber that is 8 feet long. If the ground is soft around the area protect the side from which you will be coming in and out the most with a ¾" thick sheet of plywood. A piece about 2 feet by 4 feet will usually be sufficient. Use a larger piece if you find it necessary. You will, most likely, be moving the soil that you remove from the pit, with a wheelbarrow. A wheelbarrow full of soil is quite heavy and can leave a rut in the ground as you go back and forth. If this is the case and you don't want to have to fill in the rut later, cover the path over which you will be wheeling your wheelbarrow with lengths of 2 foot wide plywood sheet.

As you dig deeper, check the sides of the pit with a plumb bob or level to make sure that the walls are vertical. If you run into rocks embedded in the earth, work around them if possible. If they are too large you will have to find some way to remove them in order to proceed.

4.1.3 Hold Back the Dirt

If your soil is sandy and loose, you may need to set up some boards to prevent the walls from crumbling into your pit. See the picture on the next page. Even if you soil is firm you will need to make sure that the walls do not crumble as you dig. Use plywood boards to hold the dirt back until you have reached a depth of 6 feet. If you have rainy days during the digging process you will have to set up a temporary canopy using a 12 foot by 16 foot plastic tarp to keep the area dry. It is not fun if your pit fills up with rain water.

4.1.4 Provide Ample Drainage

Speaking of rain, I found that I ran into an especially rainy season. Even though I had my pit covered the rain saturated the ground to the point that water started to seep in through the walls. I decided that a trench around the perimeter of the floor of the pit would be essential. You would be well advised to do the same. Dig a trench measuring about 1 foot deep and 1 foot wide. Fill the trench with 2 inch rock. Choose hard rock like granite or river rock rather than limestone which can crumble over time. You can

get rock at a variety of places like a gardening store, a hardware store or a store that sells landscaping supplies. You will have to check the prices to find which is the most economical. Landscaping stores usually supply rock in bulk rather than in bags. If you have a pickup truck, this may be the cheapest way to get your rock. Eventually you are going to need almost 4 tons, so do a bit of calling around to get the best price.

After you have filled the trench with rock cover the sides of the pit with landscape fabric. See the picture on the next page. Landscape fabric is available at your local hardware store or gardening supply store. Get the heaviest kind available, it will last longer. Also get a bag of landscape fabric spikes. These are plastic spikes about 6 inches long designed to keep landscape fabric in place. Use the spikes to hold the landscape fabric against the walls of your pit. It will keep dirt from washing into the pit during a rainy spell.

4.1.5 Construct the Floor and Walls

You are now ready to construct the floor and walls of the reservoir. For this you will need treated lumber that is able to withstand wetness and rotting. Use lumber that has not been treated with arsenic. This type of treated lumber is always clearly marked with a tag stamped as ACQ and does cost more but is environmentally sound. Sometimes the Q is printed on the tag looking like the number 2 (like this - *Q*). ACQ lumber comes in different grades as shown below. Be sure to buy the kind that is able to withstand contact with fresh water.

* Above Ground (.25)
* Ground Contact (.40)
* Fresh Water Contact (.40)

ACQ is a copper based preservative system that is the most cost-effective alternative to CCA. ACQ has been researched and tested since the late 1980's. It was introduced commercially in 1992. ACQ provides dependable performance and is building code compliant. It treats a wider range of softwood lumber species, and is quality assured by a third party (TPI or SPIB). Like CCA, ACQ lumber is treated against decay and insect damage. The Southern Pine Inspection Bureau (SPIB) is a non-profit corporation that writes grading rules for Southern Pine and is dedicated to the maintenance of high standards in the Southern Pine lumber industry. The Truss Plate Institute (TPI) is an agency whose mission is to establish methods of design and construction for trusses in accordance with the American National Standards Institute's accredited consensus procedures for coordination and development of American National Standards

In early 2004, retail lumber dealers in the United States, began phasing in ACQ (Alkaline Copper Quaternary) lumber to replace CCA (Chromated Copper Arsenate) lumber. The Preservative manufacturers are voluntarily taking this action based on negative public perception of CCA, media coverage and growing consumer interest in using an alternate wood preservative. Chromated copper arsenate (CCA) is a chemical wood preservative containing chromium, copper and arsenic. It has been used since the 1940s in pressure treated wood to protect wood from rotting due to insects and microbial agents. EPA has classified CCA as a restricted use product, for use only by certified pesticide applicators.

ACQ is ideal for structural uses, sill plates, outdoor furniture, playground equipment, patios, decks, garden edging and landscaping structures. The preservatives in ACQ products meet American Wood Preservers Association standards.

The chemicals used in ACQ will corrode ordinary galvanized fasteners, therefore special consideration must be taken when working with ACQ lumber. Hot dipped or stainless steel fasteners MUST be used with ACQ . The two most important things to remember when selecting deck fasteners, framing nails, decking nails or screws are: (1) holding capacity, and (2) resistance to corrosion. Using the wrong fasteners can compromise the appearance, longevity, and safety of an outdoor project. Many fasteners traditionally used with treated lumber are no longer recommended for this application. Check with your lumber supplier for the correct fasteners.

The above information is available on the internet at:
http://www.mamsco.com

Construct a frame with the lumber. It will rest inside the pit. The outside dimensions should be 5½ feet in every direction. A drawing of the frame design is shown for your reference. The frame needs only three walls. The fourth side is the outside of the basement wall. After constructing the frame you will attach ½ inch thick cement board on the three sides and the base with cement board screws every 8 inches.

The next step is similar to taping sheetrock with joint compound. Use 3 inch wide fiber tape and a compound meant for sealing joints in cement board. Not sure what fiber tape is? Check with an employee in the Building Materials department of your local hardware store. The compound for sealing joints in cement boards comes as a dry powder in bags and has to be mixed with water according to the instructions on the bag. Use a mason's trowel to apply the mixture to the seams. Smooth it out with a 10 inch wide joint compound blade sometimes called a joint knife. There are other tools in this department that you should consider purchasing since they will make your job easier, like a corner tool which is particularly useful for seams at right angles to each other and a square trowel used for smoothing out large flat areas. Repair any holes in your basement wall with mortar mix to prevent any possible water seepage. Give the entire assembly at least 24 hours to dry properly. You may need more time if the weather is damp.

After the enclosure is dry, you will need to reinforce the space between the cement board and the dirt. Get lengths of ½ inch steel rebar. Cut 18 lengths 5 feet each and 27 lengths 18 inches each. Using wire meant for tying rebar, tie 2 of the 5 foot lengths to 3 of the 18 inch lengths. Tie the remaining lengths the same way. You will have 9 such assemblies, 3 for each of the sides of the enclosure next to the dirt. You do not need any for the remaining side against the basement wall. Drop these assemblies, 3 per side, evenly spaced, into the space between the dirt and the cement board enclosure. They should stay up without any additional support. Next, get about 30 bags of concrete mix. You may need more depending on your situation. Mix about 2 bags at a time with just enough water to make a stiff mixture that is still fluid enough to pour into the space between the cement board enclosure and the dirt. Mix and pour the concrete until it is level with the top of the cement board. The poured cement and rebar will give sufficient strength to hold the dirt around the enclosure from shifting. Give it about 2 or 3 days to harden properly.

4.1.6 Insulate the Floor and Walls

When the cement board joints and basement wall are dry and hard, it is time to add insulation. There is a great likelihood of the inside of the reservoir getting wet during excessively rainy weather. For this reason it is a good idea to use foam board insulation instead of fiberglass. Fiberglass is a wonderful insulator when dry, but useless when wet. Foam board offers excellent moisture resistance so it won't lose any insulating power, even after long-term exposure to wet soil and ponding water. Its rugged construction readily withstands the pressure of backfilling and resists the effects of acids and other decaying elements in the soil.

Foam board insulation comes in various sizes. There is white foam board which looks just like the styrofoam that is used for packing stereo equipment. Then there is the pink kind made by Owens Corning. The pink one has a higher R value than the white one. Get the pink kind. For this project you will need several boards measuring 4 feet by 8 feet and 2 inches thick. These boards sometimes come pre-scored for use in basements between studs that are 16 inches apart. These usually cost more than un-scored boards. Get the kind that's cheaper in your area. The 2 inch thick boards have an R value of 10.

R-value is the measure of resistance to heat flow. Insulation materials have tiny pockets of trapped air. These pockets resist the transfer of heat through material. The ability of insulation to slow the transfer of heat is measured in R-values. The higher the R-value, the better the insulation material's ability to resist the flow of heat through it.

Since heat rises and the soil temperature below the frost line remains relatively constant we are more concerned with heat loss closer to the surface of the ground. Consequently you need to use only 2 inches of insulation on the floor of the pit. Use 4 inches for the walls, (two boards doubled up) and eventually we'll make a cover using 3 boards giving us a total thickness of 6 inches. Working with the insulation is easy. Measure and cut the pieces on a large flat surface using an ordinary utility knife. If you mark and score both sides of the board, you will find that it breaks along the scored line quite easily and cleanly. Use foam board adhesive such as PL300 to bond the edges and the boards to each other. Foam board adhesive is specially designed not to eat into the foam when applied. Buy the correct adhesive in tubes and a caulking gun to match. At my hardware store these items are available in the Paint department.

After cutting the foam boards, attach the bottom to the cement board floor with foam board adhesive. Next attach the walls. Since the walls use 2 layers of insulation, attach the second layer to the first using the foam board adhesive. Finally, run a bead of adhesive along the corners to make a good seal. The adhesive develops a skin within a few minutes but does not dry properly for at least 24 hours.

4.1.7 Install a Sump Pump

I learned the hard way that in spite of having a drainage ditch around the perimeter of my pit, I still needed a way to remove the water that seeped in through the soil during the unusually rainy weather at the time I built my reservoir. Consequently, I highly recommend installing a sump pump. The pump you get needs to be small enough to slide into a PVC pipe 8 to 10 inches in diameter. The one I found was a Flotec®, ¼ HP pump designed for surfaces that collect water such as basements, rooftops, window wells or any shallow flooded area. It is small enough to fit inside an 8-inch diameter PVC pipe and has sufficient pumping power to remove water deeper than 3/16 inch. It includes a 15 foot power cord and automatic on/off capability.

The 8 inch PVC pipe usually comes in 10 foot lengths. Sometimes the hardware store will also have pre-cut pieces 5 feet long. In this case get a 5 foot length.

Flotec® ¼ HP pump

[handwritten note:] Homesun – Solar power on internet. [google]

) foot length of PVC pipe 1 inch in diameter and the fittings
to the output of the pump. The Flotec® pump I used, has a
male thread. The fittings to connect this to a 1 inch pipe are readily

rill bit, drill 8 holes, evenly spaced around the 8 inch PVC pipe
one end. Then, using a hacksaw cut the pipe so that you have
s at the end of the pipe. This end will rest on the bottom of the
reservoir in one corner. The semicircular notches will allow any water accumulation to
flow to the inside of the pipe. The pump will detect any water deeper than 3/16 inch.

At the corner where you install the 8 inch PVC pipe drill a half inch hole with a
concrete drill bit so that it is approximately below the sump pump. This will allow any
remaining water to drain into the soil.

Hold the 8 inch PVC pipe vertically in place using ¾ inch metal strap available
in the plumbing department. Cut a length of strap long enough to go around the PVC
pipe and attach the ends to the walls of the enclosure using 6 inch concrete screws. The
screws have to be long enough to go through 4 inches of insulation and into the cement
board. If the concrete you poured behind the cement board is too hard for the screw,
you may need to use a concrete bit to drill a pilot hole first. Make sure the power cord
from the pump comes over the top of the 8 inch PVC pipe.

4.2 Construct the Rock Enclosure

Now it's time to make the steel enclosure that will hold the rock for the reservoir. A number of materials will work as storage media in home, farm or small business solar heating systems; but only three are generally recommended at this time: rock, water (or water / antifreeze mixtures) or a phase-change chemical substance called Glauber's salt. These are the materials that most consistently meet the criteria for selecting a storage medium: namely, the ability (1) to deliver heat to its application points at a desirable temperature, and (2) to do it cheaply, based not so much on cost of the material as on cost of the total system and its maintenance. I did not choose Glauber's salt due to its cost and availability.

Water as a storage material has the advantages of being inexpensive and readily available, of having excellent heat transfer characteristics, and of being compatible with existing hot water systems. Its major drawbacks include difficulties with system corrosion and leakage, and more expensive construction costs.

As a storage material, rocks are cheap and readily available, have good heat transfer characteristics with air (the transfer medium) at low velocities, and act as their own heat exchanger. Main disadvantages are their high volume-per-BTU-stored ratio compared to water and phase-change materials (which means a bigger heat storage area), and difficulties with water condensation and microbial activity. If the dew point of the air coming into the storage is higher than the rock temperature, the moisture in the air condenses on the rocks. Moisture and heat in the rock bed can lead to microbial growth.

Rock storage is the most reliable of the three storage systems because of its simplicity. Once the system is installed, maintenance is minimal and few things can decrease the performance of the storage.

Air solar collectors are usually used with rock storage devices. Since air collectors are cheaper and more maintenance-free than liquid collectors, a system using rock storage and air solar collectors seems the most logical for residential heating. However, other circumstances, such as the availability of cheap materials, limited collector or storage space, or incompatibility with the existing heating system, may dictate the use of a water or phase-change material storage device. Remember, however, that the ultimate deciding factor should be the initial and maintenance costs of the system.

Although size of rock selected will be determined primarily upon cost, in general, the larger the size the better for storage purposes. The main reason is that it takes less power to force the heat transfer air through large stones than through small ones. Rocks less than an inch in diameter are too small; whereas ones more than 4-6 inches in diameter are too large, because of insufficient heat transfer surface area.

If gathering your own rock for storage, look for roundish field stones in the 2- to 4-inch diameter range. But don't be overly concerned about size; settle for a 2-inch septic gravel if you'd have to pay a premium for larger rock. If available, old house brick is a good storage material when stacked to permit air flow.

More important than rock size is uniformity of size. If there is too much variation, the smaller stones will fill in the voids between the larger stones, thus increasing the air blower power requirement. Also, avoid those types of rock that tend to scale and flake, such as limestone. The resulting "dust" is picked up by the heat transfer air and becomes a problem to remove if you install an air circulating fan.

If you plan to blow air through the rock bed, it's necessary to know the amount of power needed. In general, the faster the air flow and/or the smaller the rock size, the greater the power requirement. I have not shown it in this design but if you choose to force the air flow you should consider a solar powered fan to avoid using more energy than you are trying to save by building your solar heating system. In the design here convection due to the difference in temperature at the top of the reservoir versus the bottom will cause sufficient air flow through the spaces between the rocks. Over a period of time the temperature will equalize to a large extent. There will constantly be a slight shift in temperature as heat is drawn off by the water preheater, but for the most part the temperature will remain relatively stable.

In the design used in this book I decided to use 2 inch river rock which is available through landscaping supply companies. These companies will usually deliver the rock for an additional charge. If you happen to own a pickup truck capable of hauling one ton at a time or more and don't mind the extra work you can haul the rock yourself, although you should calculate to decide which option is cheaper for you.

4.2.1 The Steel Frame

A steel enclosure is needed to contain the rock. You should construct this above ground where you have plenty of space to move around. The enclosure is constructed so that there is a few inches of space on the outside to allow air flow much the same as a convection oven. In a convection oven the part that holds the racks has a space all around it. The burner below heats the air which then flows around the chamber to allow it to be uniformly heated. The bottom of the reservoir will have a coil of ¾ inch copper tube through which a mixture of propylene glycol and water that has been heated by the solar collector, will flow. The air in contact with this tube will be heated, and, because heat rises, the hot air will flow upwards through the spaces between the rocks heating them up. Once the air reaches the top it flows around the outside of the steel enclosure down to the bottom where it passes through vents at the bottom which allow it to come in contact with the hot copper coil and the process repeats itself. As the air circulates in this manner, it heats the rocks. The rocks retain the absorbed heat until it is drawn off to heat water or your house.

The frame for the enclosure is made using steel angles. The best place to buy steel is from your local steel yard. I find that the pre-cut pieces sold in hardware stores winds up being quite expensive. The steel at the steel yard usually comes in 20 foot lengths. They will often cut it in half at no charge if you buy the whole length. Steel is sold by weight not by length. This is why it is usually cheaper to buy it at a steel yard. You can buy aluminum and other metals there as well. You will need a metal chop saw to cut your steel. If you don't have one, you should buy one. They are not very expensive and you will be able to cut your steel more accurately than the steel yard.

For the frame you will need:

1½" x 1½" x 1/8" – 4 pieces 4' 6" each
1" x 1" x 1/8" – 16 pieces 4' 8" each
¾" x ¾" x 1/8" – 12 pieces 4' 8" each

Construct the frame using the above pieces as shown in the picture. The 4 pieces of 1½" x 1½" x 1/8" will form the four corners of the enclosure. The 16 pieces of 1" x 1" x 1/8" provide support around the four sides. The ¾" x ¾" x 1/8" pieces are used at the bottom to form a vent. Keep the spacing between these pieces no more than ¾" from the bottom of one to the top of the other to prevent rocks from falling through the spaces. Refer to the diagram for what the finished frame should look like. You can decide for yourself whether to weld the frame or use nuts and bolts like I did. If you have welding equipment, you may prefer to weld the pieces. Either way, you need to make sure that the joints are strong.

4.2.2 Attach a Sheet Metal Exterior

Once the frame is assembled, sheet metal is attached to create the steel enclosure. Use 22 gauge galvanized sheet steel. This is sufficient to withstand the outward pressure from the rocks when filled. 22 gauge galvanized steel is 0.0336 inches in thickness. You can use thicker steel sheet if you want but the weight becomes difficult to manage unless you have enough helpers when you are ready to drop the enclosure in the pit. Using thicker steel also means it will cost more since steel is sold by weight.

The dimensions for each piece of sheet metal are determined by the dimensions of your frame. Attach the sheet steel so that it is flush with the top of the frame. This will leave 6" at the bottom for venting. The vents are formed by the ¾" x ¾" angle steel spaced ¾" apart at the bottom of the frame. When filled, the rocks will not spill out from between the spaces but will let air flow through.

When your steel enclosure is assembled you will need help to get it into the pit. Have two helpers inside the pit to guide and support the frame as it is lowered. You will need at least two more helpers to carry the frame over to the pit and lower it in. Position the frame against the basement wall so that the two adjacent sides have an equal amount of space between the frame and the insulation. See the picture below for reference.

4.3 Construct the Heating Coils

We are now ready to construct the heating coils. This is relatively easy. You can find 60' coils of ¾" copper pipe at your hardware store. Large hardware stores which have a separate plumbing department will usually have such items.

4.3.1 Type L Soft Copper Tubing

Copper piping is durable and comes in a variety of grades. Type M is the most basic grade and meets minimum building codes. It is thin-walled tubing and is marked for identification with red lettering. Type L is about twice as strong as type M and is a good choice for upgrading plumbing systems. It is marked with blue lettering. Type K is even stronger and is intended mainly for commercial use. It is marked with orange lettering. Soft copper tubing is sold in a coil and bends easily. It's well suited for use in tight spaces and in situations involving awkward bends, and it requires fewer joints.

Because soft copper tubing is sold in coils, it is the ideal choice for our purposes. Make sure you double check and get type L. Type M will be easily crushed under the weight of the rock which will be piled on top of it.

Regardless of grade, all copper pipe accepts the same fittings. Copper fittings may include tees, male threaded adapters, female threaded adapters, couplings, 45-degree offsets and 90-degree elbows. Depending on your situation you will some or all of these fittings.

Refer to Chapter 3 for a brief description regarding sweating copper joints. Once again you are cautioned that sweating copper joints properly takes practice and care. Improper use of the propane torch can cause fires and great personal injury. If you are unsure, get help or practice in a safe environment until you are confident. Additional tips are available on the Do-It-Yourselfer's web site. They are worth checking out. The web address is **http://www.diynetwork.com/diy/pl_pipes_fittings**. Look under the section titled **Pipe Basics**.

4.3.2 Soldering the Pipes and Fittings

Refer to the illustration on the previous page regarding assembling the coils. The picture shows two 60 ft coils one on top of the other. You will be assembling three pairs of coils. It is most convenient if you place one coil on a flat surface like a cement slab and lay the second coil so that the inner end of the coil below is opposite the inner end of the coil on top. In this configuration the outer ends of the coils will be on opposite sides of the assembly and much easier to manage for connecting to the pipes that will be connected to them for entry into the basement wall which will be done a bit later.

Use a ¾" copper coupler to join the inner ends of the two coils. You may find that you have to trim off a small section of the end which may not be perfectly round due to damage in shipping. In this case you will not be able to get the coupling onto the end of the pipe. After trimming the end with a tubing cutter, make sure that you check by dry-fitting the ends. Sand the ends with emery cloth, fine sandpaper or a wire brush designed for this purpose. Also clean the inside of the fitting. Apply tinning flux designed for lead-free solder. Push the two ends of the pipes so that they meet inside the fitting. Heat the pipe ends with the propane torch until the solder melts when you touch the pipe with it. It should flow freely around the joint. Remove the torch and wipe off any drips with a damp rag.

Assemble the other two pairs of coils in the same manner. Put these assemblies aside for now. You will need them a bit later.

4.4 Electronic Controls for Your System

In order to control the heating system some method is needed for turning the pumps that circulate the antifreeze on and off. As described in Chapter 2, the system works by collecting solar heat in the collector panel. The pipes in the collector panel are filled with a mixture of non-toxic propylene glycol and water. This mixture is circulated through copper tubes down into the copper heating coil located in the bottom of the heat reservoir pit. Circulating pumps installed inline with the copper tubes are used to circulate the antifreeze mixture. The pumps will be installed inside the basement for maintenance and accessability. To prevent the pumps from running when the temperature of the collector panel drops at night to a temperature that is lower than the temperature in the reservoir pit, we have to install a controller that senses the temperature difference and turns the pump off. A sun detector circuit can also be added to remove power from the system after sundown and before sunrise. Having a sun detector in addition to the Differential Thermostat will take care of times during the day when clouds might cover the sun for an extended period of time.

If you understand electronic schematics and have some knowledge of electronic assembly, a detailed description of assembling such a controller is given here. If you do not know how and would prefer to get a fully assembled controller, you may write to:

Kenneth Clive
2936 34th Avenue S
Minneapolis, MN 55406-1708

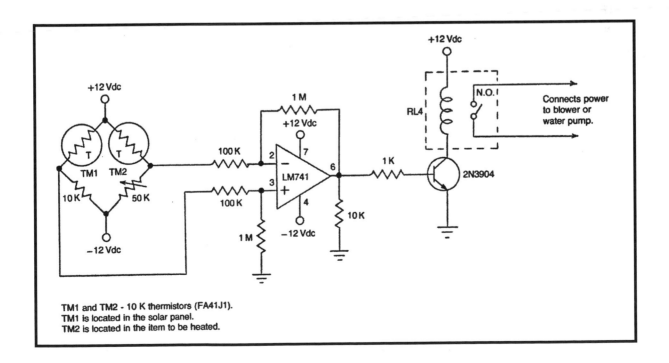

TM1 and TM2 - 10 K thermistors (FA41J1).
TM1 is located in the solar panel.
TM2 is located in the item to be heated.

Include a check or money order of US $75.00 payable to Kenneth Clive. This price is subject to change at any time due to prices of components. You will be asked to pay the difference before it is shipped. In your order mention that you are requesting a Differential Thermostat for Solar Heating System.

4.4.1 Build a Differential Thermostat

The schematic for the differential thermostat is shown on the previous page. All the resistors are ¼ Watt carbon resistors. The 50K potentiometer is also ¼ Watt.

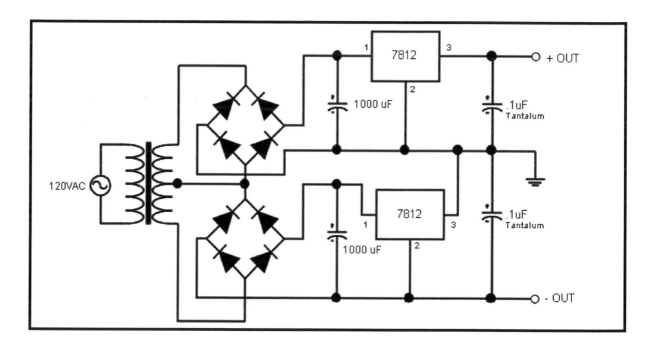

The relay RL4 is a 12V relay whose contacts are normally open and handle 2 amps of current. The thermistors TM1 and TM2 are 10K NTC types. These are commonly available at electronics parts stores like Radio Shack®. A power supply capable of supplying 500mA is needed as well and is shown above. Notice that the differential thermostat requires both positive and negative 12VDC. The schematic above shows how this is achieved. Use a 24VAC center tapped transformer capable of 500mA and two bridge rectifiers connected as shown. The 1000 μF capacitors are 15V electrolytic. Be sure to observe proper polarity. The .1 μF capacitors are Tantalum and are also polarized.

The sun detector circuit shown on the next page can be used to remove power

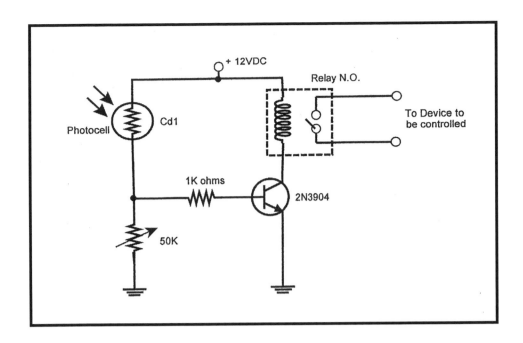

from the system when the sun's radiation is not present. The relay switch should be normally open when solar radiation is absent and closed when solar radiation is present. If you are unsure of the open and closed position of the relay switch test it with a multimeter set to the ohms position. Depending on the kind of multimeter you have, the closed switch position will give you a reading of zero similar to touching the probes of the meter together. The open switch position will give you a reading of infinity, similar to when the probes are not touching each other.

The printed circuit board for the differential thermostat is shown on the next page for reference as to component layout. If the components you purchase, such as the capacitors, rectifier, and relay, are of a different size you will need to modify the printed circuit board accordingly. Sanford Corporation®, the makers of Sharpie® markers, have an industrial super permanent marker available at office supply stores like Office Max which resists processing chemicals and cleaners. Their item number is **13763** which comes 3

to a pack. This is a good etch-resistant marker to use for creating printed circuit boards. Like all markers keep the cap on when not in use to prevent the marker from drying out.

Copper clad boards are available at electronics parts stores like Radio Shack®. Get the kind of boards that have copper on only one side. They also have kits with etching

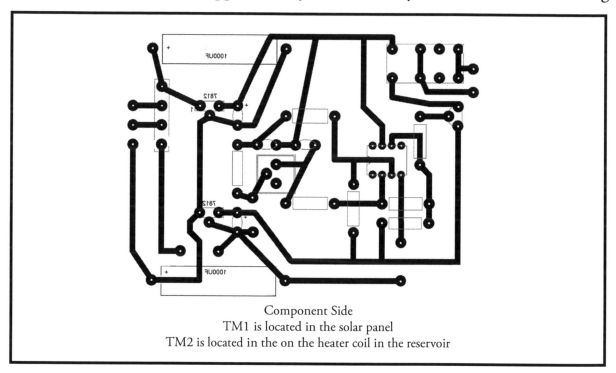

Component Side
TM1 is located in the solar panel
TM2 is located in the on the heater coil in the reservoir

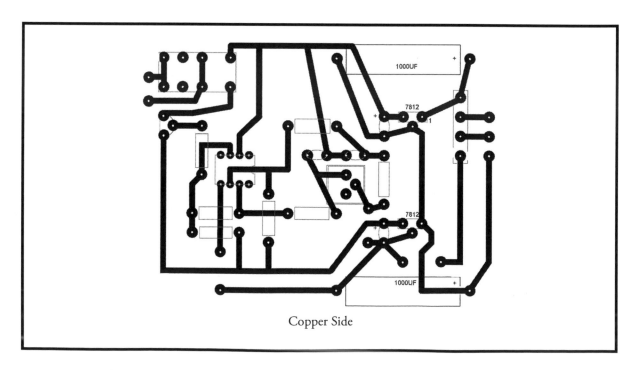

Copper Side

solution and etch resistant markers. Ferric Chloride is commonly used as the etching solution and is extremely corrosive. Follow all the warnings that come with it. Wear old clothes and chemical resistant gloves in case you should happen to splash the solution onto yourself. Chemical resistant gloves are available at most hardware stores. Ordinary latex gloves meant for kitchen use are a lot cheaper and can be used if you do not suffer from latex allergies. Stains from ferric chloride are impossible to remove and will eat through the fabric and anything with which it comes in contact so it is extremely important to avoid splashing.

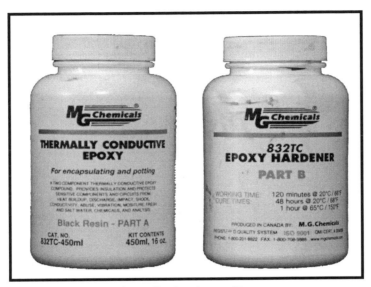

Thermally Conductive Epoxy

When your circuit board is assembled install it in a project box. This will protect the components and also protect you from accidentally coming in contact with high voltage.

4.4.2 The Differential Thermostat Sensors

TM1 and TM2 are the sensors that will detect the difference in temperature between the collector panel and the reservoir pit. In order to do this they need to be in thermal contact with the copper tube in the collector and the copper heating coil in the reservoir pit. This is accomplished by inserting the thermistor inside ¼" copper tube 1½" long, and holding it in with thermally conductive epoxy. Thermally conductive epoxy is available from **Allied Electronics®**. Their toll-free number is **1-800-433-5700** and in their 2007 catalog it is part number **661-3530**. The manufacturer's part number is **832TC-450ML**.

The epoxy mentioned above takes 2 days to harden so you should put your assembly somewhere where it will not be disturbed for the required amount of time. You can help the hardening process by warming the assembly to around 75 degrees F. Be careful not to heat it too much otherwise bubbles will form in the epoxy. The wire leads of the thermistor should be insulated so as not to come in contact with the inside of the copper tube. Pull an appropriate length of insulation off some 22 gauge copper bell wire and thread it onto the leads of the thermistor leaving about ½" bare wire at the end. 22 gauge wires will be soldered to these ends, one pair being long enough to reach from the collector panel down to where the controller is located in the basement and the other from the heating coil in the reservoir pit through the basement wall to the controller. Save these sensors till after the copper heating coil is installed in the reservoir pit.

4.4.3 Monitoring the Temperature

The temperature should be monitored at various points to give you an idea of how well your system is working. You can do this using some relatively inexpensive electronic temperature displays. These units have a remote probe attached to wires connected to the display. I got mine from Johnstone Supply. Their catalog number is B12-777. They have an online catalog at www.johnstonesupply.com. The wires may not be long enough to reach from the collector panel or the heater coil in the reservoir to where you have the displays mounted in the basement. Simply cut off the wires and add a sufficient length of 22 gauge thermostat wire to make up the difference. You will need four of these remote thermometers for this project. One probe will be located in the collector panel on the roof and the other three will be located on the copper pipe coils in the reservoir pit. Later in this chapter you will read more about the pipe coils.

4.5 Through the Basement Wall

Blocks used to build basement walls come in different sizes depending on when the house was built. The blocks used in my basement wall measure 24" x 9" x 12". The 12" dimension is the thickness of the wall. Depending on the condition of the basement wall, it may take a considerable amount of work to remove a block. Since the length of the block I removed was 24" it gave me sufficient room to pass 6 copper pipes and a ½" electrical conduit evenly spaced through this opening.

4.5.1 Take Out a Block

Use a cold concrete chisel and a 3 pound hammer to chisel out a block from the basement wall. Before you start, use the carpenter's adage "Measure twice cut once". You don't want to make a mistake here. If you are unsure of whether the block(s) you are planning to remove are structurally critical to the rest of your house, you should get expert advice from an architect or construction engineer. You may need to install steel jack posts temporarily inside your basement until you can replace the block(s) you remove with a steel insert. Jack posts are available in most major hardware stores like Home Depot. They come in different lengths and are usually adjustable for height. You should measure the height of your basement from the floor to the joists to determine what size you need to purchase. Install the jack posts according to the manufacturer's recommendation.

4.5.2 Replace the Block With a Steel Support

Every precaution must be taken to prevent losing the structural integrity of your basement wall. Remember, your entire house is resting on it and weakening one spot may be detrimental to the rest of the support structure. If you installed steel jack posts in the previous step, leave them in place at this time.

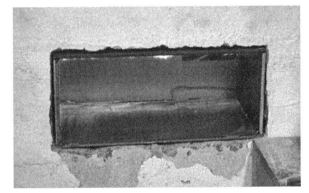

Steel support installed

Measure the dimensions of the opening you created in your basement wall very precisely. Your measurements should not be off by more than 1/16". Using these dimensions, have your steel yard build you a steel insert using 3/8" steel plate. They can sometimes do it while you wait. The plate steel should be measured and cut on one of their industrial shears. See the illustration for reference. These facilities usually have a welding shop as well. Once the four pieces of plate steel are cut the welding shop can weld them together for you. There is no need to have the weld run the entire length of the joins. Three or four good tacks with a TIG or MIG welder are sufficient. Make sure that the person doing the welding checks to make sure that each corner is square. The entire assembly will weigh quite a bit so, unless you are used to lifting more than 40 pounds, get someone to assist you.

This assembly should be placed in the space where the block(s) was/were removed and should fit snugly. If there is any need to use a hammer to get the assembly in place, make sure to use a wood block so that the hammer does not damage it. Once in place, you can remove the steel jack posts supporting your basement ceiling.

Depending on the thickness of your basement wall, cut a sufficient number of pieces of the same foam insulation which was used to insulate the reservoir pit, to insert in the steel frame you just installed. Mark each piece using the plywood boards with the holes for the copper pipes and electrical conduit and drill holes through the foam. It is much easier to do this before inserting the foam in the steel frame so that all the holes line up properly.

4.5.3 The Interface Between the Outside and Inside

Cut two pieces of ½" thick plywood measuring 30" x 15". Mark these pieces and drill holes as shown in the illustration. Two holes at the ends will allow you to pass 3/8" threaded rods to hold the pieces of plywood firmly against the outside and inside of the basement wall. The other holes will allow the copper pipes and electric conduit to be passed through. It is best to measure the marks for these holes and drill them carefully. It becomes very difficult to line them up if you don't. You may have to alter the hole placement to suit your situation if your blocks are of a different size. Give yourself at least 4" between each copper pipe. If your basement wall is 12" thick, cut the copper pipes to a length of 20". This way when you have to solder fittings at each end you won't burn the plywood.

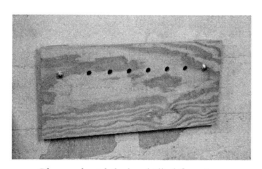

Plywood with holes drilled for pipes

4.5.4 Install the Pipes and Electrical Conduit

Cut six pieces of ¾" Type L copper pipe long enough to extend through the basement wall, plywood boards and protrude about 3 inches on either side. My basement wall is 12" thick so I cut my pipes 20" long to extend beyond the plywood on the inside and outside of the wall. Cut a piece of ¾" electrical conduit to the same length as the copper pipes you just cut. Send the copper pipes and the electrical conduit through the holes in the plywood and the foam insulation. You may need to play with the alignment of the holes in the plywood and foam insulation to feed the pipes through the wall. You can make the job a bit easier by marking the foam and pre-drilling before inserting it inside the steel support. The copper pipes and electrical conduit will be secured to the plywood backing both on the inside and outside of the basement wall, and make up the interface between the reservoir and the plumbing which will run through the basement and up to the solar collector on the roof.

4.6 Filling the Rock Enclosure

It is time now to fill the rock enclosure. This will be done in a few stages rather than all at once since the heating coil has to be located towards the bottom. Next will be the coil that will transfer heat to assist in heating the house and thirdly will be the coil that connects to the water heater to preheat the water coming in from the city water supply.

4.6.1 Choosing the Rock

Some information has already been given earlier in Section 4.2 regarding the choice of rock to use for this project. I recommend using river rock measuring between 2 to 4 inches. This will allow adequate air flow to heat the reservoir uniformly.

4.6.2 The First Stage

Start by putting in enough rock to come above the level of the vents in the steel enclosure. Level this out so that the surface of this layer is as uniform as possible. Once you are satisfied with the leveling of the rocks, place a piece of expanded steel measuring 48" x 48" over it. You can get expanded steel at the steel yard. See the illustration. It usually comes in 4 ft. x 8 ft. sheets and in a few different sizes of hole spacings. You can get the sheets cut at the steel yard. Choose sheets where the spaces are small enough to prevent the rocks from falling through. Purchase 3 sheets and have them cut in half so you will have 6 pieces all together. One sheet will go

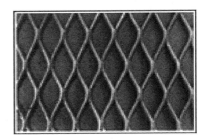

Expanded Steel

below each copper pipe coil and one above it. The expanded steel will allow air to pass through while protecting the copper coils from getting punctured by any sharp rocks. It will also equalize the pressure on the coil to prevent it from being crushed by the weight of the rest of the rocks that will be poured on top.

Hose clamp

4.6.3 The Heating Coil

The first of the three copper coil assemblies can now be laid in the reservoir on top of the expanded steel mesh. Center the coil in the enclosure and solder extensions using lengths of straight copper pipe to reach up to the basement wall. Next using a combination of copper fittings and lengths of copper pipe connect the coil to the pipes going through the basement wall. See the illustration. Be careful not to burn the plywood

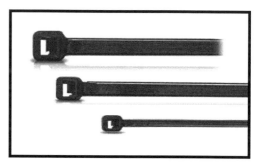

Cable ties come in different sizes.
Choose ones that are about 8" long.

backing when soldering the copper fittings. Mark the plywood to indicate that the pipes are connected to the heating coil.

Using a 1" hose clamp attach a thermometer and differential thermostat probe that you made earlier to the center of the coil and bring the wires towards the basement wall. Use nylon cable ties to attach the wires along the pipes running vertically up against the basement wall and thread the ends through the electrical conduit into the basement. You can find cable ties in the electrical department of hardware stores. Attach identifying tags to the wires so you know where they are coming from. Also identify the wires so you know which pair is for the thermostat and which is for the thermometer. Cover the pipe coil with another 4ft x 4ft square of expanded steel.

The picture on the previous page shows how I connected the pipe coils before I discovered that it was easier to lay the two coils as shown on page 43.

The first coil covered with rocks

Second stage coil inside the reservoir. Note the expanded steel mesh over the rocks and under the coil. Also notice the PVC pipe in the top left corner for the sump pump.

4.6.4 The Second Stage and Coil

Now you can fill more rock over the expanded steel mesh until the enclosure is full to about half way up. Once again, level out the rocks carefully and lay down another sheet of expanded steel. Place another copper pipe coil assembly and as before using a combination of fittings and straight copper pipe connect the ends of this coil to two pipes going through the basement wall. This coil will absorb heat which will be used to heat the house.

Use a 1" hose clamp to attach a thermometer probe to the center of this coil. Mark the wire with an identifying tag before sending it through the electrical conduit into the basement. As before tie the wire to the vertical copper pipe with nylon cable ties to prevent it from flopping around. Cover the pipe coil with a 4ft x 4ft square of expanded steel.

Fill more rocks over the previous coil until the enclosure is about ¾ full. As before, level out the rocks, lay down a sheet of expanded steel and place the third coil of copper pipe. Connect the ends of the coil with lengths of straight copper pipe and fittings to the two remaining copper pipes going through the basement wall. Attach a thermometer probe to the center of the coil. Tag the wires of the probe to identify them and pass them through the electrical conduit making sure to first secure the wire against a vertical copper pipe. Cover the pipe coil with a sheet of expanded steel as before.

Third stage coil. The black wire is connected to the thermometer probe and attached to the copper coil.

4.6.5 The Third Stage and Coil

The third stage is for pre-heating water coming from the city supply before it goes to your water heater. The idea here is to have the water heater not work so hard. If you get lots of bright sunny days you may develop enough heat in your reservoir that the water heater may not come on except on rare occasions. Since the copper coil is 120 feet long and ¾" in diameter, it holds approximately 2¾ gallons of water.

For those of you interested in knowing how this is calculated, here's the formula: The volume of a cylinder is $\pi r^2 l$, where r is the radius of the pipe and l is the length. The diameter is .75" and the length in inches is 1440. π is approximately 3.14159. These values give us 636.1725 cubic inches, or 0.368 cubic feet. There are 7.4805 gallons to one cubic foot. So 0.368 x 7.4805 gives us 2.75 gallons.

Depending on the rate at which hot water is drawn from your water heater, and the temperature of your reservoir, the water entering your water heater will be warm and your water heater will not come on as often as before.

4.6.6 Finish Filling the Rock Enclosure

The rock enclosure can now be filled with enough rock so that the level of rock is a few inches below the top of the enclosure. Pass an electrical cord through the conduit which has an in-line socket on the end. This will allow you to connect the power cord for the sump pump. This cord can simply lie on the surface of the rock. This completes the filling of the reservoir.

The reservoir covered with a wooden deck. The surrounding area was then
properly finished with a brick surround and grass was replanted.

4.7 Insulating and Covering up the Reservoir

Finally the top of the reservoir should be covered with sufficient insulation so that almost none of the heat may escape. I recommend using 3 layers of 2" thick foam insulation. This will give a combined R value of 30. Measure the dimensions of your reservoir wall to wall. Measure from the outside edge of the foam wall on one side to the outside edge of the foam wall on the opposite side. Reduce this dimension by about a ½". This will be the dimension you will use for the foam cover. Cut the pieces and stack them in an overlapping manner to cover the top without any gaps. You will need to cut a hole for the PVC pipe coming vertically out of the sump pump. Do a dry fit to make sure your pieces fit properly. When you are satisfied, glue all the pieces for the cover together using foam adhesive such as PL300.

Purchase insulation tape measuring 1/8" thick and 2" wide in a roll long enough to go around the perimeter of your reservoir twice. This tape is similar to the kind used to seal leaky windows and doors in winter. Get the kind that has an adhesive backing. This tape needs to be attached to the top edge of the foam wall. This will fill any gaps due to uneven surfaces where you cut the foam board to form the wall. If your wall is excessively uneven try to get ¼" thick insulation tape. The adhesive holding the foam cover should be properly set in about 24 hours. Make 4 holes in the foam cover and thread ¼" rope through them. Tie ½" dowels about 4" long on either side of the cover. This will enable you, with the help of an assistant to place the cover over your reservoir. It will also be useful if you ever need to remove the cover for maintenance.

Next, measure the reservoir from the outside edges of the concrete wall that surrounds the foam insulation. Construct a wooden frame to support a deck-like structure which will be the final cover for your reservoir. Build it so it will be sturdy enough for people to walk over without hurting anything in the reservoir. This completes the construction of the reservoir.

Chapter 5

Getting Ready To Mount The Collector

5 | Getting Ready To Mount The Collector

In order to mount the collector panel on your roof you will need a rack to support it. The rack in turn must be secured to the roof so that it will remain attached in very strong winds. There are some factors regarding the rack which must be considered. We will try to keep this project as simple as possible. Obviously, this means that some sacrifices will have to be made in terms of maximum solar gain.

5.1 Fixed or Moveable Mounting Rack

The first decision I made regarding the mounting of the panel was whether to make it fixed or moveable. What does that have to do with anything, you ask. Well, you don't need me to tell you that the sun rises in the east and sets in the west. What this means in terms of solar collection is that for maximum performance the panel should always be perpendicular to the rays of the sun. Not only does the panel need to be oriented towards the sun as it moves across the sky daily, it also needs to change its orientation with the seasons. The sun is at its highest during the summer solstice around the 21st of June and at its lowest during the winter solstice around the 21st of December. The only way the panel can be oriented perpendicular to the sun's rays at all times is to move it by means of some form of electronically controlled motorization. While this is not impossible, it is fairly difficult to construct, and is beyond the scope of this book.

5.2 The Optimum Angle For Your Collector

So what's a do-it-yourselfer to do? Unless you have a strong aptitude for electronics and mechanical design you'll want to stick with the simpler option here.

A basic understanding of the geometrical relationship between the earth and the sun is necessary here. Imagine a carousel and pretend that the sun is at the center of the carousel and that the horses are the planets. Every time one of these "planets" goes around once, is the same as the earth going once around the sun. The earth takes approximately 365 days to go around the sun once. While the earth is revolving around the sun, it is also rotating thus causing day and night. The imaginary line around which the earth rotates is its axis. If the axis were perfectly perpendicular to an imaginary line drawn from the center of the sun to the center of the earth, all of our days and nights would be the same number of hours all year round. However, this is not the case. The axis of the earth, instead of being perfectly perpendicular to that imaginary line drawn from the center of the sun to the center of the earth, is tilted at an angle of approximately 23.5°. This gives us the changing seasons, and also the reason we see the sun higher in the sky during summer and lower in the sky during winter. Of course, this is only true for people living in the northern hemisphere. For people living in the southern hemisphere the tilt of the earth's axis causes them to experience winter when the northern hemisphere is experiencing summer and vice versa. In the southern hemisphere, the sun appears lowest in the sky during the same period that the sun appears highest in the sky in the northern hemisphere.

To help people like us make some calculations for projects such as this one, geographers have come up with some imaginary lines circling the surface of the earth known as latitude lines. The equator is the latitude line above which we call the northern hemisphere and below which we call the southern hemisphere. The latitude

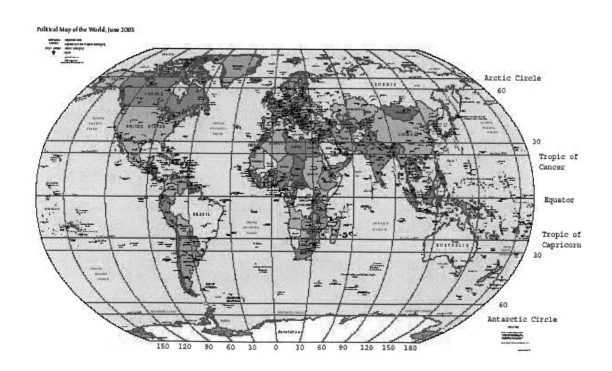

lines are measured in degrees starting at the equator which is considered 0°, and the North Pole which is considered 90°. Minneapolis, Minnesota, where I live, is almost exactly at 45°. If you consult a world atlas you will be able to find the latitude of the place where you live. Latitude lines for the southern hemisphere also start with 0° at the equator and increase to 90° at the South Pole.

This basic information helps us to choose the angle for our solar collector. If we were to choose the same angle as the latitude of our location we would have a panel perfectly perpendicular to the sun's rays on the day of the spring and fall equinox. On these two days, supposing that both days were completely cloudless, the solar panel would absorb the most solar radiation. The radiation absorbed on all other days would be less due to the sun's rays striking the panel less directly. The 2 days when the suns rays would strike the solar panel least directly would be the days of the summer and winter solstice. While it sounds like a good idea to orient the solar collector so that its angle is the same as the latitude, it is a better idea to choose an angle that takes advantage of the sun's angle during the colder months. It is during this period that more heat is needed.

The calculation to obtain this angle is as follows:

A commonly used mathematical symbol for an unknown angle is the Greek letter Θ (theta). Without going into all the trigonometric calculations, the optimum angle $\Theta_{(opt)}$, measured from the horizontal for the latitude at your location, represented by L is

$$\Theta_{(opt)} = 90° - (L + 16.58°)$$

For latitudes in the southern hemisphere $\Theta_{(opt)}$ is calculated as

$$\Theta_{(opt)} = 90° - (L - 16.58°)$$

In my case L = 45°, so my optimum angle came out to be

$$\Theta_{(opt)} = 90° - (45° + 16.58°)$$
$$\Theta_{(opt)} = 90° - (61.58°)$$
$$\Theta_{(opt)} = 28.42°$$

This angle represents only the angle from the local horizontal to a line perpendicular to the surface of the collector. The collector should also be pointed towards the south if you live in the northern hemisphere, or north if you live in the southern hemisphere. You should know that the magnetic north and south are slightly different from the geographic north and south. The best way to determine the geographic south (if you live in the northern hemisphere) is to place a straight stick in the ground, perfectly vertically and to trace the shadow it casts at exactly 12:00 noon. This shadow points to the geographic north. The opposite direction is south. If you live in the southern hemisphere a vertical stick in the ground will cast a shadow pointing due south at noon.

Instead of repeating instructions for both hemispheres, from now on angles and directions will be for the north, above the equator. Just reverse them if you live in the south, below the equator.

Appendix A has a table showing Latitude angles for several major cities in the United States. If your city is not in the list, you may also find it at:

http://www.bcca.org/misc/qiblih/latlong_us.html

In the table online you will find the latitude and longitude in degrees and minutes. If you wish you can convert the minutes to a decimal value by dividing by 60. You may find this helpful when doing your own calculations. My opinion as to how closely you calculate, based on the latitude of your location, is that you can round up to the nearest whole degree without losing any significant solar absorption. This is because you will be building a fixed mounting rack, and also because the latitude values given are for very specific locations, so if you're off by a tenth of a degree in your calculations, it really doesn't matter. You may also be interested to know that the axis around which the earth rotates does not remain fixed, but wobbles slightly over a period of about 435 days. This was discovered in the 19th century and is named the Chandler wobble after its discoverer.

5.3 Building a Fixed Rack for Optimum Winter Conditions

So far we know the optimum angle of orientation for the collector for winter. We must now calculate the angles for a rack that will hold the solar collector panel at this angle. If you live in North America, your home, most likely, has a sloping roof. If one of the slopes of your roof is south-facing, your job is relatively easy. If your house does not have a south facing roof slope, or, if you have a flat roof, the problem is compounded due to the fact that you have to figure out a way to mount and anchor the rack in a safe manner. Remember that if you mount anything on your roof you should take into account the added weight of the panel and its ability to withstand strong wind gusts. Consult your local city offices for any applicable building codes you should comply with, and any suggestions they may have for your project.

That being said, this book gives you enough information to make any modifications you may need to customize the design to your specific application. Also keep in mind that your "south-facing roof" may not be true south. However, as long as it is within ± 5° of true south you should be just fine. It is a good idea to check this carefully before building your mounting rack. From here on we will assume you have a south-facing roof.

In order to determine the angles for your mounting rack, you need to know the angle or pitch of your roof. The pitch of your roof is the amount of rise per horizontal distance converted to an angle. Converting rise per horizontal distance to an angle requires trigonometry. Alternatively, once you have determined the rise per horizontal distance, you can consult the table in Appendix B to determine the angle. House builders use similar tables during the construction of new houses. Booklets containing many useful conversion tables are often available at a hardware store.

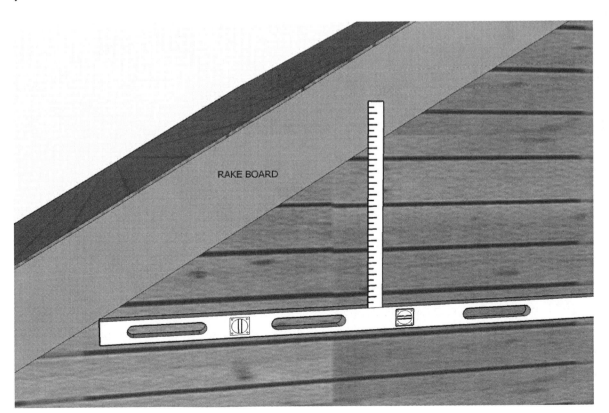

RAKE BOARD

The pitch of your roof is most easily measured against the "rake board". The rake board is the board along the top edge of your roof that forms an upside-down V. See the illustration on page 69. Using a carpenter's level and a ruler we can easily find the pitch. Mark off 24 inches from the end of the carpenter's level with a piece of tape. Then holding the carpenter's level horizontal with the end against the lower edge of the rake board, hold the ruler vertically at the 24 inch mark and measure the number of inches from the carpenter's level to the bottom of the rake board. Use these measurements to find from the table in Appendix B what the angle of your roof is. Subtract the angle of your roof from the optimum winter angle for your collector to get the angle for your roof rack.

The following example is to illustrate how I did my calculations. You will need to do your own calculations for your project. The angle of my roof turned out to be 34°. As I mentioned earlier the latitude for my location is 45° and so the optimum winter angle for my collector is 28.42°. In other words, my collector needs to be at an angle such that a line perpendicular to its surface makes an angle of 28.42° with the horizontal. To find this angle I subtracted 28.42° from 90° which is 61.58°.

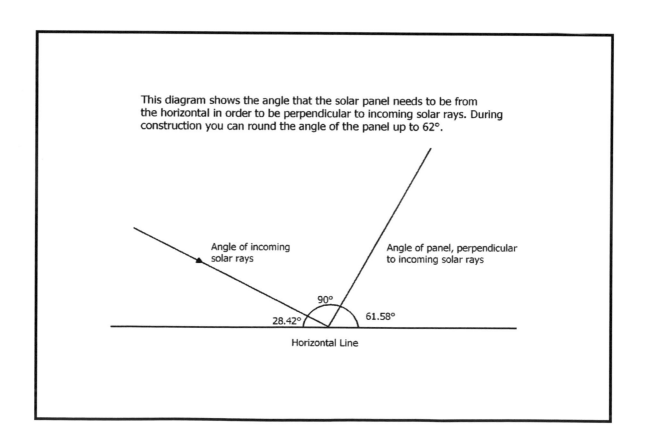

This diagram shows the angle that the solar panel needs to be from the horizontal in order to be perpendicular to incoming solar rays. During construction you can round the angle of the panel up to 62°.

Angle of incoming solar rays

Angle of panel, perpendicular to incoming solar rays

90°

28.42°   61.58°

Horizontal Line

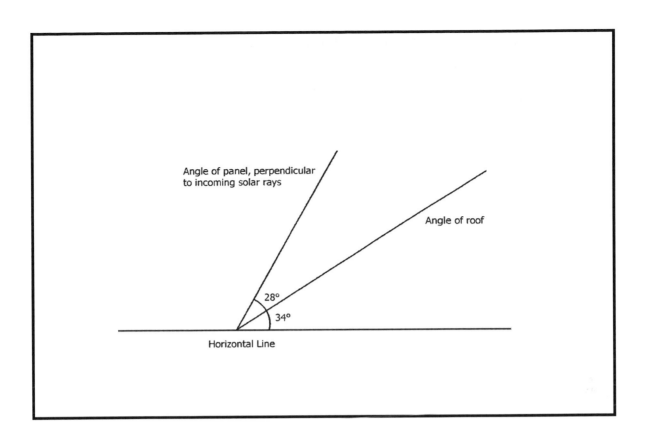

Since my roof angle is 34°, I subtracted 34° from 61.58° to get the angle for my rack which turns out to be 27.58°. I can round this off to 28°. Next, I made a triangular template from ¼ inch plywood to use to check the angle of the rack. Using the table in Appendix B, I found that 28° is a rise of 13 inches in 24. I drew a right angled triangle on the plywood using a carpenter's square. I marked off 24 inches for the base and 13 inches at right angles to the base. I joined the two end points with a straight edge to form the triangle and cut it out carefully with a hand saw. Once again, these measurements were for *my* project and are shown here as an example. Yours will be different unless your latitude and roof angle is the same as mine.

To build the rack you will need the following materials:

1.5" x 3" Rectangular aluminum tube 1/8" thick
1.5" x 1.5" Aluminum angle 1/8" thick
2" x 2" Aluminum angle 1/8" thick
3" x 3" Aluminum angle 1/8"   thick

¼ x 20 x 1.25" galvanized hex head bolts with flat washers, lock washers and nuts

¼ x 20 x 1" galvanized hex head bolts with flat washers, lock washers and nuts
¼ x 2" lag screws

(The lag screw shown in the picture is not actual size.)

3/8" x 1" galvanized hex head bolts with flat washers, lock washers and nuts
3/8" x 2 ½" galvanized hex head bolts with flat washers, lock washers and nuts

The pictures on the next few pages show how you will construct the rack using the materials above. The rack is triangular in cross-section and I have referred to the members of the triangle by its orientation to the surface of the roof where it will be installed. For example, I refer to the piece made from the rectangular aluminum tube as the "bottom member" because it forms the bottom edge of the triangle and will lie flush on the surface of the roof. The purpose of the bottom member is to provide a certain amount of space between the panel and the surface of the roof to allow rain and leaves from nearby trees to slide down the roof easily and prevent any debris from getting trapped under the panel. The drawings follow a logical sequence so that you will be able to follow the steps to constructing the rack.

Referring back to Chapter 3, the external dimensions of the solar panel are 30" x 48" x 5 ½". The rack will consist of triangular sections assembled together. Each of the triangular sections will be a right angled triangle. If we choose the hypotenuse of the triangle to be 48", and knowing that the angle at the base needs to be 28°, we can use the plywood template cut out earlier to measure 28°, along with a carpenter's square to mark off the lengths we need for the triangular section.

You can also calculate the lengths easily if you are comfortable with a bit of elementary trigonometry. If the base is denoted by the letter **x** and the perpendicular by the letter **y** then using 48" for the hypotenuse and 28° for the angle needed in addition to the roof angle,

$$\frac{x}{48} = \cos 28°$$

$$x = 48(\cos 28°)$$

$$x = 48(0.8829)$$

$$x = 42.3814"$$

$$x = 42\tfrac{3}{8}"$$

$$\frac{y}{48} = \sin 28°$$

$$y = 48(\sin 28°)$$

$$y = 48(0.4694)$$

$$y = 22.5346"$$

$$y = 22\tfrac{1}{2}"$$

The base of the triangle is made with 1.5" x 3" rectangular aluminum tube, the side perpendicular to the base is made with 1½" x 1½" angle aluminum and the third side is made with 2" x 2" angle aluminum. See the illustration.

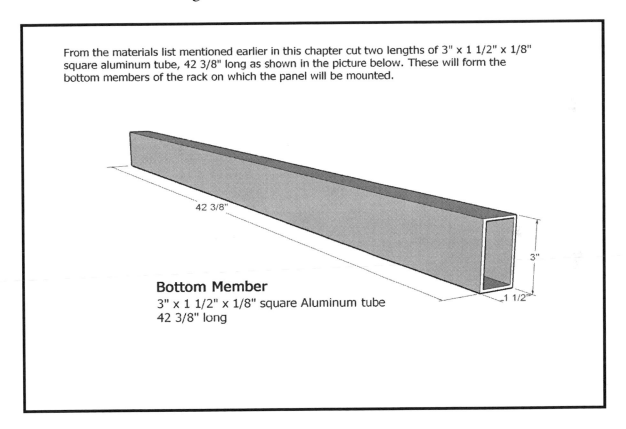

From the materials list mentioned earlier in this chapter cut two lengths of 3" x 1 1/2" x 1/8" square aluminum tube, 42 3/8" long as shown in the picture below. These will form the bottom members of the rack on which the panel will be mounted.

42 3/8"

3"

1 1/2"

Bottom Member
3" x 1 1/2" x 1/8" square Aluminum tube
42 3/8" long

Two lengths of 3" x 3" angle aluminum along the top and bottom edge of the triangular rack will eventually hold the panel securely on the rack.

The series of illustrations below show in a step-by-step manner how the complete rack is assembled. You will start by making two triangular assemblies based on the measurements calculated. The two triangular assemblies have been referred to as the left and right assembly. They are identical in their measurements except that they are assembled as mirror images of each other. You may wonder why the vertical member in the assembly is 25½" instead of 22½". The reason is that 3" are added to take the height of the bottom member into account.

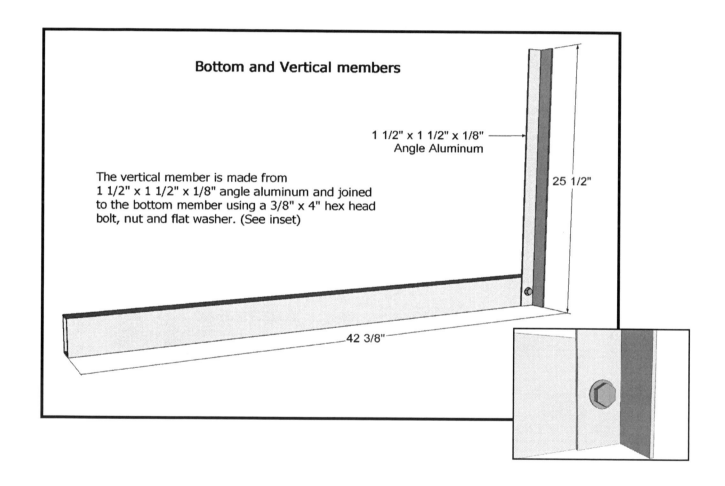

Bottom and Vertical members

1 1/2" x 1 1/2" x 1/8"
Angle Aluminum

The vertical member is made from
1 1/2" x 1 1/2" x 1/8" angle aluminum and joined
to the bottom member using a 3/8" x 4" hex head
bolt, nut and flat washer. (See inset)

25 1/2"

42 3/8"

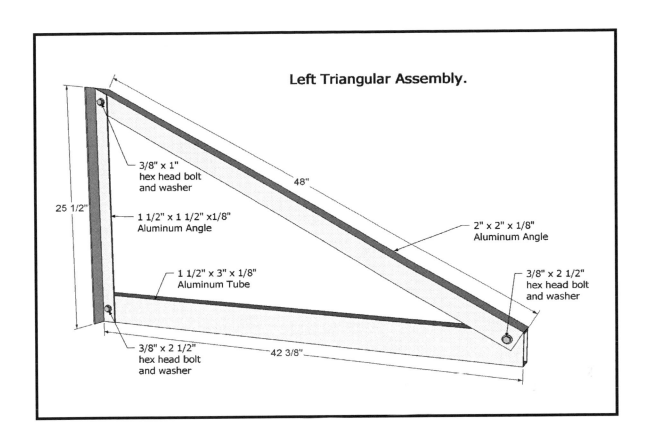

Left Triangular Assembly.

3/8" x 1"
hex head bolt
and washer

25 1/2"

1 1/2" x 1 1/2" x1/8"
Aluminum Angle

48"

2" x 2" x 1/8"
Aluminum Angle

1 1/2" x 3" x 1/8"
Aluminum Tube

3/8" x 2 1/2"
hex head bolt
and washer

3/8" x 2 1/2"
hex head bolt
and washer

42 3/8"

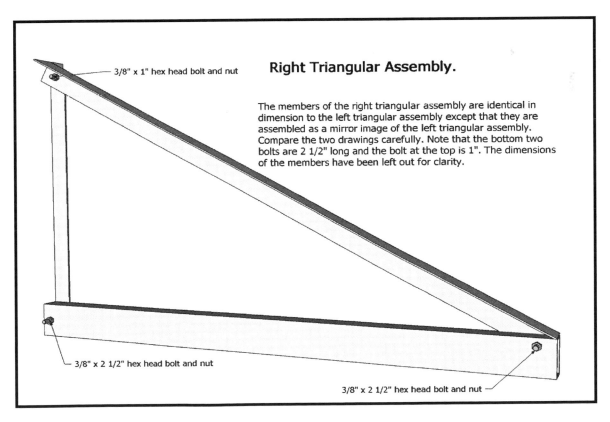

3/8" x 1" hex head bolt and nut

Right Triangular Assembly.

The members of the right triangular assembly are identical in
dimension to the left triangular assembly except that they are
assembled as a mirror image of the left triangular assembly.
Compare the two drawings carefully. Note that the bottom two
bolts are 2 1/2" long and the bolt at the top is 1". The dimensions
of the members have been left out for clarity.

3/8" x 2 1/2" hex head bolt and nut

3/8" x 2 1/2" hex head bolt and nut

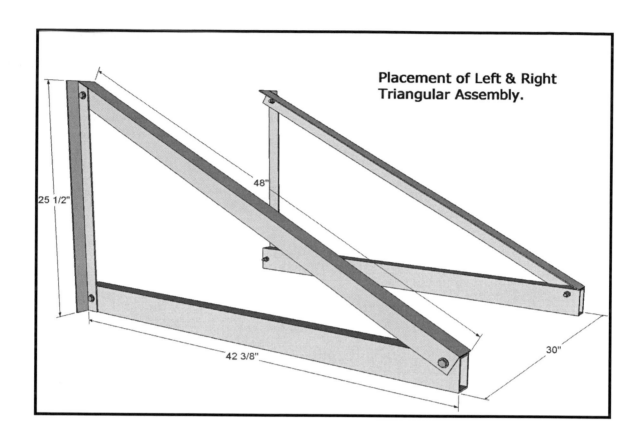

Placement of Left & Right Triangular Assembly.

25 1/2"

48"

42 3/8"

30"

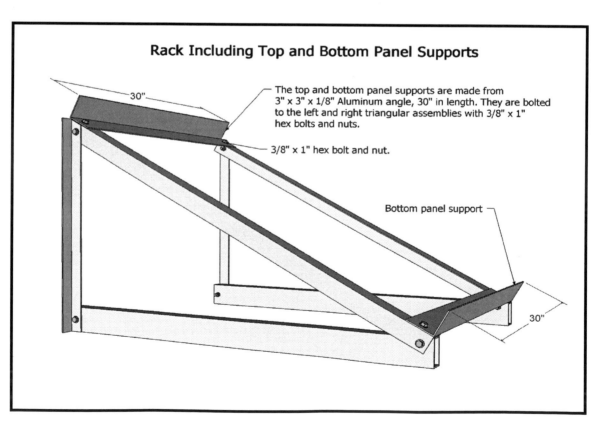

Rack Including Top and Bottom Panel Supports

30"

The top and bottom panel supports are made from 3" x 3" x 1/8" Aluminum angle, 30" in length. They are bolted to the left and right triangular assemblies with 3/8" x 1" hex bolts and nuts.

3/8" x 1" hex bolt and nut.

Bottom panel support

30"

Rack Showing Back Support

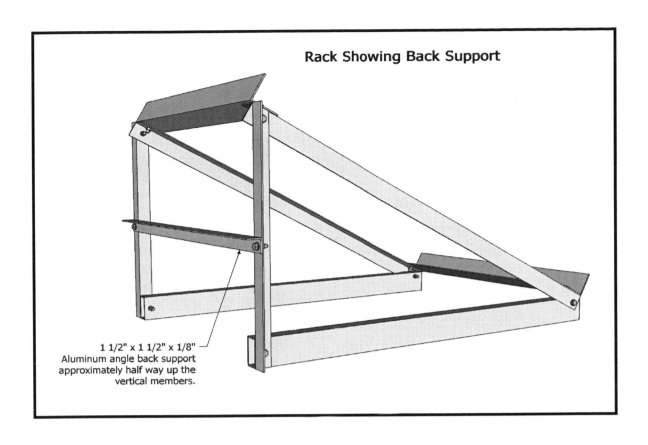

1 1/2" x 1 1/2" x 1/8"
Aluminum angle back support
approximately half way up the
vertical members.

Rack Assembly With Mounting Angles

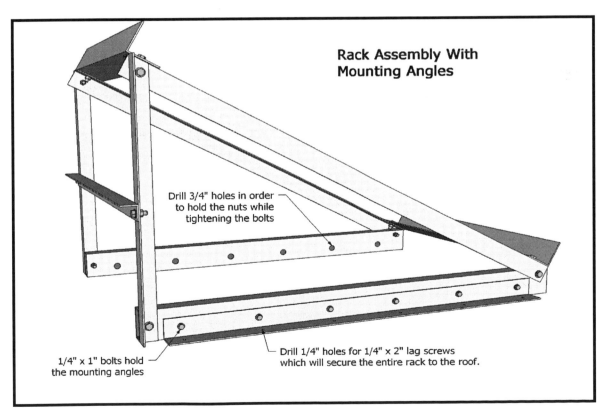

Drill 3/4" holes in order
to hold the nuts while
tightening the bolts

1/4" x 1" bolts hold
the mounting angles

Drill 1/4" holes for 1/4" x 2" lag screws
which will secure the entire rack to the roof.

5.4 Installing the Rack on Your Roof

Once the rack is complete you should be able to mount it on your roof. In the materials list I have suggested that you use ¼ x 2" lag screws. If you have multiple layers of shingles on your roof you will need to adjust the length of the lag screws accordingly. If the lag screws do not penetrate the roof boards beyond the shingles you run the very real danger of having your entire installation coming off the roof.

Set up a couple of extension ladders so that the tops of the ladders are about a foot above the edge of the roof. If the pitch of the roof is too steep to allow a good foothold, use a roofing ladder. Tie a rope securely to the rack and with the help of an assistant on the ground, pull the rack up along the extension ladders onto the roof until you have it in the proper position. The "proper position" will depend on your unique situation and should be the place where you get the most sun for the longest period each day. Hold the rack in position temporarily with a couple of 16 penny nails through the holes you drilled for the lag screws. Pre-drill pilot holes where the lag screws will go and install the lag screws.

5.5 Working Safely on the Roof

It is extremely important to follow some safety precautions while working on the roof. Working on a sloping surface means that you have to balance yourself along with your tools and supplies while managing the parts of your rack and panel. It can be very frustrating, not to mention quite dangerous, when tools inadvertently slip out of

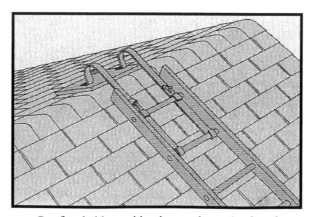

Roofing ladder and hook over the peak of roof

your hands and slide down the sloping roof possibly hitting someone standing on the ground below. Using ladder hooks to hold an extension ladder or using a roofing ladder will help to climb up and down the slope of the roof safely. Ladder hooks and roofing ladders have a curve at the top end which allows you to hook them over the peak of the roof to prevent the ladders from sliding down. Use a wooden board under the ends of the hooks will spread the weight and prevent damage to shingles.

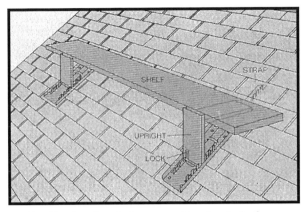

Roofing brackets help to provide a level platform

In addition to the ladders you should use adjustable metal brackets designed to make a level platform for tools and supplies. See the illustration above. Your brackets may be slightly different. These brackets have a shelf that support a 2-by-10 plank and are available in the roofing supplies section of hardware stores. The bracket is held on the roof with ordinary nails. Be sure to wear rubber-soled shoes with a good grip to keep from slipping and damaging the shingles. Use cordless power tools to avoid tripping over electric extension cords. Do not work on the roof in rainy weather.

Chapter 6

Install the Collector Panel

6 | Install the Collector Panel

It is now time to install the collector panel on the rack which was previously mounted on the roof. Be sure to observe proper safety procedures as mentioned in the last chapter. The panel is bulky and heavy and you can undergo severe personal injury if you are not careful. In addition, if you do not have a secure grip on the rope being used to pull up the panel, it may fall and become badly damaged, or worse, hurt someone on the ground below.

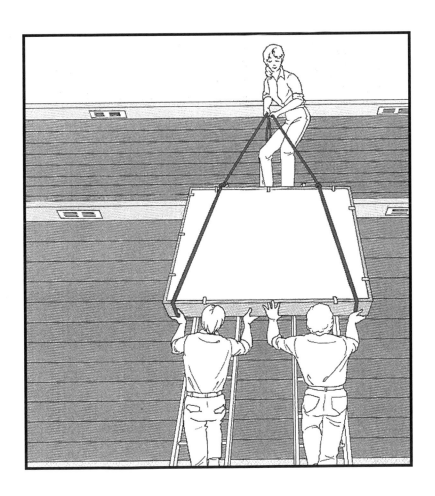

6.1 Hoisting the Panel Onto the Rack

You may come up with your own way of hoisting the panel onto the roof and onto the rack. However you do it depends, in part, on your individual situation. Keep safety in the forefront whatever you do. In my case, I installed a pair of hand operated winches on the peak of my roof and pulled up the panel by turning the crank a few turns alternately while my neighbor observed from the ground below to let me know if the panel was coming up without tipping to one side or the other. Once the bottom edge of the panel cleared the lower lip of the rack it was safe to remove the ropes and seat the panel properly in the rack. I found that it helps considerably to use a couple of extension ladders as runners along which to pull the panel as shown in the illustration on the previous page. This method allows a fair amount of the weight of the panel to be supported by the ladders while pulling the panel up..

6.2 Securing the Panel on the Rack

Drill holes for some #10 x 1" galvanized sheet metal screws no more than 6" apart along the top and bottom supports of the rack and screw the panel in place. The weight of the panel and the screws through the rack into the walls of the panel will be sufficient to hold it even in a very strong wind if you have secured the rack as described previously. The panel on my roof has withstood wind gusts of up to 40 m.p.h.

6.3 Attach the Thermometer and Differential Thermostat Probes

After securing the panel on the rack, attach a thermometer and the differential thermostat probe for the panel to the copper pipe with a metal hose clamp. Drill a ¼" hole through the side of the panel and thread the wires from the probes through the hole to the outside of the panel. Use a plastic project box from Radio Shack® like the one shown here, (Catalog #: 270-1801) to make a junction box for the wires from the panel.

You will be installing the glass cover next so this would be a good time to make sure that the inlet and outlet pipes exit the panel properly and in a manner that allows you to make the necessary connections to the pipes coming up through your roof. Double-check everything and then check it once more.

Project Box Radio Shack® Catalog # 270-1801

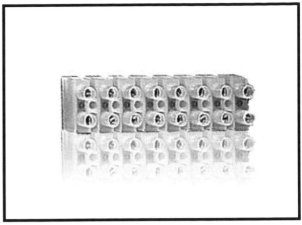

Use a terminal strip like this one to connect wires from
the solar panel to wires from the basement.

6.4 Installing the Glass Cover

Once the thermometer and thermostat probes have been attached and the wires
threaded through the side of the panel to the outside you can pull up the glass or acrylic
cover in the same way as you pulled up the panel. Be sure to protect the surfaces of the
glass to prevent it from getting scratched or damaged in any way while you are pulling it
up. One way to do this is to tape strips of cardboard on the side of the glass which will
ride up the ladders you are using as rails. The glass should be placed in the recess formed
by the dadoes of the panel walls. Use silicone caulk to seal the edges around the glass
to prevent any moisture from getting in. A word of caution, before you seal the glass in
place make sure that you have completed everything that requires you to get inside the
panel. Once you have sealed it, it will be extremely difficult to remove the glass for any
purpose.

Chapter 7

Install the Rest of the Plumbing

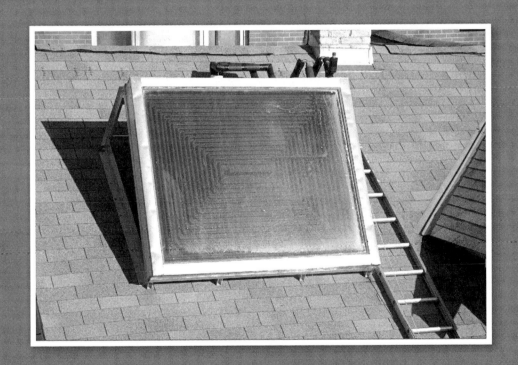

7 | Install the Rest of the Plumbing

Installation of the rest of the plumbing depends on your individual home structure so the information provided here is general in nature. After reading this, if you feel that it is beyond your capability and outside your comfort zone, you should get professional help. As I have mentioned several times already, safety is a primary consideration.

7.1 Connect the Pipes

In Chapter 3, you saw several pipe fittings which are needed throughout this project. You will be using some or all of these fittings and perhaps, others not shown. The attendants at most hardware stores are able to help you find what you need for your situation. The system may seem complex when you look at all the pieces involved but the principle is quite simple. The part that does most of the work is a closed loop consisting of two coils of copper pipe; one in the panel on the roof and the other at the bottom of the reservoir. These two coils are simply connected to each other with a pump installed to keep the liquid in the pipe circulating. The schematic below shows the essential components of the system. Shut-off valves and other details are not shown.

At the end of Chapter 4 the copper pipes in the reservoir were passed through the basement wall inside the house. Your next job is to determine the best path to the roof so that the pipes from the bottom of the reservoir can be connected to the collector. My house is a single story house with an unfinished attic. One of the walls on the inside runs approximately down the middle of the house and has some vent pipes running through it. Vent pipes are part of the drain system and allow gases that build up to escape. These are steel pipes that protrude through the top of the roof and are around 2" in diameter. Exterior walls of houses are usually filled with insulation so you should not attempt to pass any plumbing through them. However, inside walls of houses are often hollow so that plumbing pipes and electrical wires can be passed through. It is through such a wall that you can pass the copper pipes for your heating system. If necessary, you

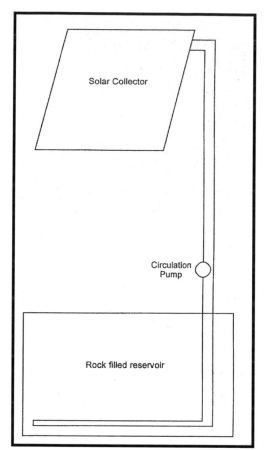

Heating System Schematic

may need to find someone knowledgeable to help you determine how to tackle this part. If you find that there is no convenient way to pass the pipes from the basement to the roof, you can run them outside the house. In such a case you should provide sufficient insulation so that the heat collected isn't lost due to exposure to the elements. Pipe insulation of varying thickness can be special-ordered through professional plumbing supply stores. These stores often sell to the public as well. It usually comes in 6 foot lengths and may or may not be split along the length. You may have to use a knife to cut along the length of the insulation to get it onto the pipe but you should tape the insulation once it is on the pipe so that heat cannot escape. I purchased mine through Johnstone Supply. It is sold under AC/AP Armaflex® Pipe Insulation. You can find the website at

www.johnstonesupply.com.

7.2 Running the Lines Through the Roof

Assuming that you are able to find a path for your pipes through an inner wall in your house you now need to pass the pipe through the roof. If your attic is unfinished the task becomes relatively easy. If your attic is finished, get expert advice regarding the options available to you. You may find that your best option is to run the pipes outside the house and simply provide sufficient insulation around them.

If your attic is unfinished, find a spot between the rafters that is closest to where you want the pipes to emerge onto the surface of the roof so that they can be connected to the pipes from the solar collector panels. You can find the spot by making a pilot hole with a ¼" drill bit from the surface of the roof and checking where it comes through the attic. Making sure to work between the rafters, use either a saber saw or a 4 ½" hole saw to cut a circular opening through the roof. Cut an 18" length of 4" diameter PVC pipe and pass it through the hole in the roof so that it is vertical. Place flashing

Photo showing the pipes coming up through the roof. The pipes are covered with ¾" thick pipe insulation. The PVC pipe is filled with an expanding foam sealant which comes in 12 oz. pressurized cans available at many hardware stores.

intended for pipes protruding from the roof, over the PVC pipe so that the boot of the flashing is centered on the PVC pipe. Push the upper half of the flashing up under the roof shingles. Remove any interfering nails with a pry bar. Fasten the lower half of the flashing over the shingles with roofing cement. If your flashing comes with a separate boot collar, slide it over the PVC pipe so that it fits snugly over the boot. If the collar is integral to the flashing just make sure that it is centered along with the flashing around the PVC pipe.

After running the copper pipes through the PVC pipe, fill the space in the PVC pipe with expanding foam sealant. Expanding foam sealant comes in pressurized 12 oz. cans available at many hardware stores. Be extremely careful when handling this product. Wear a mask that covers your nose and mouth so that you do not breathe it in. At the point where your copper pipe enters the collector panel, install a T fitting and connect a length of copper pipe and attach a coin-vent bleeder valve to allow air to escape when charging the system with antifreeze. The bleeder valve should be the highest point in the system.

7.3 Connect the Pump etc.

A circulation pump is needed in the system to circulate the antifreeze so that the heat collected by the collector panel can be transferred to the reservoir. I chose a model by Bell and Gossett model NRF-22 shown below. It comes with the flanges and gaskets

Circulator Pump

which allow you to easily connect it in line with your copper pipe. This pump is intended to be connected in a vertical section of pipe oriented as shown in the picture. If you wish, you can also find a similar pump which is designed to be Install the pump, valves and controls in the basement where the pipes come through the wall from the reservoir.

In a horizontal section of your pipe, connect an air eliminator. After the system is filled, purged and vented at high points, the air scoop will automatically separate any remaining air from the water pumped through it. The air is then automatically vented to the atmosphere. Note the arrow showing the direction of flow. The top has a tapped connection for installing an air vent.

Taco Air Scoop Model 431-6

The air vent I used was a Honeywell Braukmann model EA122A. There is an instruction sheet inside the package that details how to use this device. The air vent is opened by turning the valve body counterclockwise. This exposes the EA122A to the system. The antifreeze and air can then enter the vent chamber by opening the vent seat. The red vent cap allows air to pass through the open vent. As the air is released through the vent cap, antifreeze replaces the air in the vent chamber and a float inside rises. This closes the vent seat. When additional air enters the vent chamber, the operation repeats. The vent can be closed by turning the body clockwise isolating the unit from the system and allowing service if necessary.

The tapped connection at the bottom of the Air Scoop is provided for connecting a pressurized expansion tank. The tank shown below is a 4.4 gallon diaphragm type tank that can be pressurized to 40 P.S.I. This allows the antifreeze in the system to expand and contract as it heats and cools.

Honeywell Braukmann EA122A
Automatic Air Vent

4.4 Gallon Thermal Expansion Safety Tank

7.4 Install Shut-off Valves

Install shut-off gate valves before and after the circulation pump. It will help if the system needs maintenance at any time. In a horizontal section of the pipe install two boiler drain valves with a gate valve in the middle as shown in the sketch. This will be used later to charge the system with antifreeze.

Boiler Drain Valve

Gate Valve

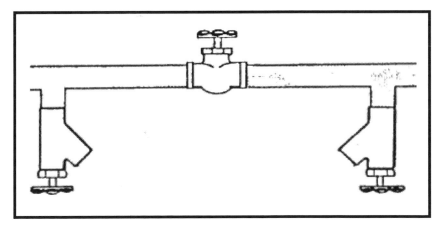

Connect the Boiler Drain valves and Gate valve as shown

7.5 Pressure Relief Valve

One additional piece, for the added safety, is a pressure relief valve. Hot water heaters are equipped with these. Install one in your system close to the expansion tank. It is fairly unlikely that your system will ever create the kind of pressure for the valve to open, but it is good to have it just in case.

Chapter 8

Connecting the Controllers, Pumps, and Thermometers

8 | Connecting the Controllers, Pumps, and Thermometers

In Chapter 6 a junction box was installed on the collector panel for wires from the thermometer and thermistor probes. Using the same path that you used to bring pipes from the basement up to the top of the roof, thread 4-conductor bell wire from the basement up to the collector panel and connect the wires to the junction box. If you drill a hole in the junction box to pass the wires, make it facing downwards and seal the hole with silicone caulk when you are done to prevent moisture from getting in. Use wire that has multicolored insulation so as to be able to differentiate the wires connected to the thermometer probe from the wires connected to the thermistor probe. This kind of wire is readily available at many hardware stores.

8.1 Wiring the Controllers

In Chapter 4 a sheet of plywood was installed inside the basement wall to attach the pipes and wires coming through. I used a 2 foot by 4 foot piece of plywood ¾" thick. It gave me enough space to clamp the pipes to support the pump and all the other devices. I mounted the thermometer displays on an 8" by 8" piece of ¼" thick piece of plywood and attached it to the plywood on the wall. I labeled each of the displays to show which temperature was being read. The picture on the next page shows one of the thermometer displays.

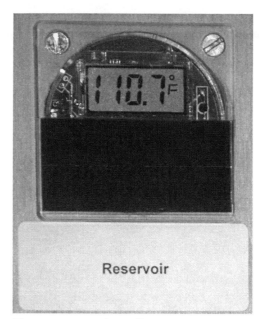

Thermometer readout of reservoir temperature

Having thermometer probes connected to various points in your system gives you a good idea of how well your system is working. The temperature above was read on a cold but sunny November morning when the outside air temperature was around 38° F. The black portion of the meter shown above is a solar cell so there is no need for batteries.

The differential thermostat and sun detector that were built in Chapter 4 may be added next. Having the sun detector is a good idea particularly since the whole point of building this project is to save energy. It would not make sense to have the pumps running on a cloudy day or after sundown since you would be taking heat from the reservoir and radiating it out to the atmosphere.

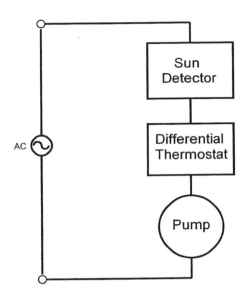

This diagram shows how the sun detector is connected to the differential thermostat and the pump which circulates the antifreeze to heat the solar heat reservoir. The sun detector shuts off power to the differential thermostat on cloudy days or after sunset. This in turn shuts off power to the pump.

Be careful when wiring the differential thermostat and sun detector since one side of the relays on both of these will be connected to 110 volts AC and you can get electrocuted if you are careless. Connect the other side of the relay on the sun detector to the power transformer and relay for the differential thermostat. Take care to insulate and cover wire junctions that carry high voltage to protect yourself as well as the electronic circuits.

The block diagram above shows how the controller circuits should be connected. The Sun Detector has a transformer that requires 110 volts AC household current to operate. Also connect the hot wire of your household AC to the relay switch on the sun detector and the Differential Thermostat. The other side of the relay switch should be connected to one side of the transformer on the Differential Thermostat. The other side of the transformer on the Differential Thermostat is connected to the neutral side of the AC supply. Connect the hot wire of the 110 volt AC household current to one side of the relay switch of the Differential Thermostat. The other side of the switch of the Differential Thermostat is connected to the black/hot wire of the circulation pump. The white wire on the pump should be connected to the neutral side of your 110 volt household AC supply. In addition, a ground wire should be connected to the body of the pump as indicated by the manufacturer. Check and recheck your wiring. You could destroy your controller circuits due to incorrect wiring. You could also electrocute yourself.

8.2 Wiring the Pumps

The controller circuits should be wired to the pump that circulates antifreeze from the collector panel on the roof to the bottommost pipe coil in the reservoir and back again. The pump comes with a complete set of instructions showing the proper wiring connections.

The second pump in the system is connected to the middle pipe coil and will circulate the antifreeze for heating your house. The wiring should be connected through a high voltage thermostat so that the pump runs only when the temperature in the area you are trying to heat falls below the temperature set on your thermostat.

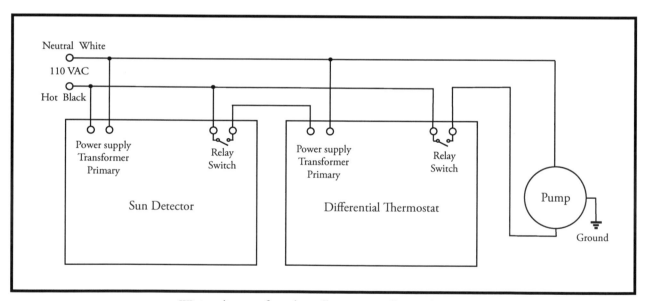

Wiring diagram for solar collector controllers and pump

Honeywell Thermostat model T498A1810

The thermostat I used for the house heating circulating pump is shown above. It is a Honeywell model T498A1810 designed for heavy electric loads such as electric heaters. It is different from the usual thermostats for furnace applications. The package wiring diagram shows it connected to an electric heater. Wire in the pump where it indicates a heater. Follow the directions that come with the package. Use a second differential thermostat to prevent the pump from running if the temperature in the reservoir falls below the temperature in the room you are heating. In this case your regular furnace will take over. Attach the probe for the differential thermostat for this part to the house heater pipe coil in the reservoir.

8.3 Charging the System With Antifreeze

Your next job is to fill the system with antifreeze. As I mentioned in an earlier chapter, get non-toxic propylene glycol. You can get this in five-gallon buckets at plumbing supply places. I have even seen one-gallon jugs at hardware stores. It is used by owners of cabins in the northern states in the winter to keep pipes from freezing.

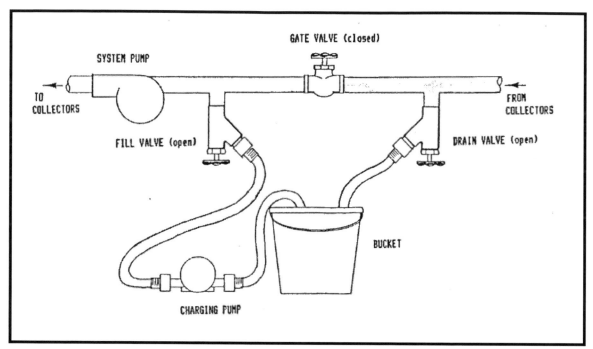

Diagram showing how the system is filled with antifreeze.

The diagram above shows how you can fill your system with antifreeze quite easily. Use a five-gallon bucket filled with a 50-50 mixture of propylene glycol and water, much the same as you would mix antifreeze for your car. See the mixture recommendations on the bucket you purchase in case you need to modify the propylene glycol to water ratio. Use clear plastic tubing so you can see what's happening.

The pump I used was a small one made by an Oklahoma company called Proven Pumps. The pump itself is a model they call a Pony Pump model 360. It is capable of moving up to 300 gallons per hour and can lift liquids from 7 feet below to 40 feet high. Fill the inlet tube using a small jug with a spout with the antifreeze mixture until you see it come through the outlet side of the pump. Put your thumb over the end of the tube and immerse it into the bucket of antifreeze mixture. The antifreeze mixture should start flowing when you start the pump. It may take a few tries to get a steady flow going. Once you have a steady flow air bubbles should start coming out of the other tube. If you seem to be having trouble, check to make sure that all the valves in your system are open except for the gate valve between the two boiler drain valves. It may also help to turn on your circulating pump.

When the system is full the antifreeze mixture will start flowing through the tube not connected to the pump and you should not see any more air bubbles. During this process you will probably need a helper to keep the bucket filled so that you don't have to keep stopping the pump to refill. My system took about 10 gallons of the antifreeze mixture. When no more air bubbles are visible, shut off both boiler drain valves and shut off the pumps. Open the gate valve between the boiler drain valves and your system should be ready to go. Up at the highest point of your piping on the roof you should have provided a bleeder valve. Check it to make sure that there is no air trapped which will prevent the circulation pump from circulating the antifreeze mixture.

8.3.1 How Much Antifreeze Do You Need

Calculate your antifreeze needs using the formula below.

1 cu ft = 7.48 gallons
$\pi r^2 l$ = cubic feet where r is the radius in feet and l is the length in feet of the copper pipe. Use 3.142 for the value of π.

Estimate the total length of copper pipe in your system by adding up the horizontal and vertical runs up to the roof, the total length used in your collector panel and the coils in the reservoir. If you made the coils in the reservoir as shown earlier, that length is 120 feet. Adding up all the lengths and inserting them in the formula will give you the number of cubic feet. Multiply the cubic feet by 7.48 to get the number of gallons you will need.

Fill the house heater part of the system in the same manner. Once again you should be able to calculate the volume of antifreeze you need using the same formula given above.

Chapter 9

Turn it on and Enjoy!

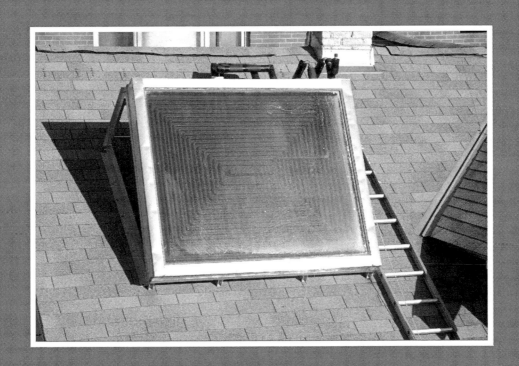

9 | Turn it on and Enjoy!

Your heating system should now be ready to operate. Inspect each section carefully to make sure everything is in order and that there are no leaks anywhere. Insulate all the pipes with the pipe insulation indicated in an earlier chapter, especially the pipes running up through an uninsulated attic and above the roof to the collector panel. Make sure that the slits that run the length of the insulation face downwards in any horizontal runs. If your pipe insulation did not come with an adhesive coating along the split, tape it up well using duct tape. Do everything you can to prevent heat loss.

On a sunny day your temperature readout for your collector panel will be quite high. I have been able to register over 200° F on sunny days during the summer when there are many more hours of sunlight than in the winter. Due to the huge volume of rocks that need to be heated, it will take many days for the temperature to stabilize so don't get discouraged by your initial temperature readings.

I have a budget plan with my gas company in order to have a predictable bill each month. In the summer months I accumulate a credit since the furnace remains off so less gas is used and in the winter when more gas is used the credit is applied to my bill. The budget amount is evaluated every year and adjusted according to my useage. Last month my budget amount was reduced to almost half my normal budget amount. Now granted, we had a pretty mild fall, but considering that the budget is adjusted for useage over an entire year, the only thing I can attribute the drastic reduction is to the addition of my solar heating system.

It is my hope that you will build your own heating system and enjoy the process and the results just as much as I have enjoyed building mine and sharing my experience with you.

Appendices

Appendix A

Latitude Angles For Several Major Cities In The United States

Note: Normally angles are measured in degrees, minutes and seconds. There are 60 seconds in a minute, and 60 minutes in a degree. In this table the minute values of the latitude of each location have been converted to a decimal value.

State	City	Latitude In Degrees
AK	Homer	59.38
AL	Birmingham	33.56
AL	Mobile	30.68
AL	Montgomery	32.38
AR	Fort Smith	35.33
AR	Little Rock	34.73
AZ	Phoenix	33.43
AZ	Prescott	34.65
AZ	Tucson	32.11
AZ	Winslow	35.01
AZ	Yuma	32.65
CA	Bakersfield	35.41
CA	China Lake	35.68
CA	Daggett	34.86
CA	Fresno	36.76
CA	Long Beach	33.81
CA	Los Angeles	33.93
CA	Mount Shasta	41.31
CA	Needles	34.60
CA	Oakland	37.81
CA	Red Bluff	40.09
CA	Sacramento	38.51
CA	San Diego	32.73
CA	San Francisco	37.61
CA	Santa Maria	34.90
CA	Sunnyvale	37.41
CO	Colorado Springs	38.81

State	City	Latitude in Degrees
CO	Denver	39.75
CO	Grand Junction	39.11
CO	Pueblo	38.30
CT	Hartford	41.73
DC	Washington-Sterling	38.95
DE	Wilmington	39.66
FL	Apalachicola	29.73
FL	Daytona Beach	29.18
FL	Jacksonville	30.50
FL	Miami	25.81
FL	Orlando	28.55
FL	Tallahassee	30.38
FL	Tampa	27.96
GA	Atlanta	33.65
GA	Augusta	33.36
GA	Macon	32.70
GA	Savannah	32.13
HI	Hilo	19.71
HI	Honolulu	21.33
HI	Lihue	21.98
IA	Burlington	40.78
IA	Des Moines	41.53
IA	Mason City	43.15
IA	Sioux City	42.40
ID	Boise	43.56
ID	Lewiston	46.38
ID	Pocatello	42.91
IL	Chicago	41.78
IL	Moline	41.45
IL	Springfield	39.83
IN	Evansville	38.05
IN	Fort Wayne	41.00
IN	Indianapolis	39.73
IN	South Bend	41.70
KS	Dodge City	37.76
KS	Topeka	39.06
KS	Wichita	37.65
KY	Lexington	38.03
KY	Louisville	38.18
LA	Baton Rouge	30.53
LA	Lake Charles	30.11
LA	New Orleans	29.98

State	City	Latitude in Degrees
LA	Shreveport	32.46
MA	Boston	42.36
MD	Baltimore	39.18
MD	Patuxent	38.28
ME	Caribou	46.86
ME	Portland	43.65
MI	Alpena	45.06
MI	Detroit	42.41
MI	Flint	42.98
MI	Grand Rapids	42.88
MI	Houghton	47.16
MI	Traverse City	44.75
MN	Duluth	46.83
MN	International Falls	48.56
MN	Minneapolis/St. Paul	44.88
MN	Rochester	43.91
MO	Columbia	38.96
MO	Kansas City	39.11
MO	Springfield	37.23
MO	St. Louis	38.75
MS	Jackson	32.31
MS	Meridian	32.33
MT	Billings	45.80
MT	Cut Bank	48.61
MT	Glasgow	48.41
MT	Great Falls	47.48
MT	Helena	46.60
MT	Lewistown	47.06
MT	Missoula	46.91
NC	Cape Hatteras	35.26
NC	Charlotte	35.21
NC	Greensboro	36.08
NC	Raleigh-Durham	35.86
ND	Bismarck	46.76
ND	Fargo	46.90
ND	Minot	48.41
NE	Grand Island	40.98
NE	North Omaha	41.30
NE	North Platte	41.13
NE	Scottsbluff	41.86
NH	Concord	43.20
NJ	Lakehurst	40.03

State	City	Latitude in Degrees
NJ	Newark	40.70
NM	Albuquerque	35.05
NM	Clayton	36.45
NM	Farmington	36.73
NM	Roswell	33.30
NM	Tucumcari	35.18
NM	Zuni	35.10
NV	Elko	40.83
NV	Ely	39.28
NV	Las Vegas	36.08
NV	Reno	39.50
NV	Tonopah	38.06
NV	Winnemucca	40.90
NY	Albany	42.75
NY	Binghamton	42.21
NY	Buffalo	42.93
NY	Massena	44.93
NY	New York City	40.78
NY	Rochester	43.11
NY	Syracuse	43.11
OH	Akron-Canton	40.91
OH	Cincinnati	39.15
OH	Cleveland	41.40
OH	Columbus	40.00
OH	Dayton	39.90
OH	Toledo	41.60
OH	Youngstown	41.27
OK	Oklahoma City	35.40
OK	Tulsa	36.20
OR	Astoria	46.15
OR	Burns	43.58
OR	Medford	42.37
OR	North Bend	43.42
OR	Pendleton	45.68
OR	Portland	45.60
OR	Salem	44.92
PA	Allentown	40.65
PA	Erie	42.08
PA	Harrisburg	40.22
PA	Philadelphia	39.88
PA	Pittsburgh	40.50
PA	Wilkes-Barre	41.33

State	City	Latitude in Degrees
RI	Providence	41.73
SC	Charleston	32.90
SC	Columbia	33.95
SC	Greenville	34.90
SD	Huron	44.38
SD	Pierre	44.38
SD	Rapid City	44.05
SD	Sioux Falls	43.57
TN	Chattanooga	35.03
TN	Knoxville	35.82
TN	Memphis	35.05
TN	Nashville	36.12
TX	Amarillo	35.23
TX	Austin	30.30
TX	Brownsville	25.90
TX	Corpus Christi	27.77
TX	Dallas	32.85
TX	El Paso	31.80
TX	Fort Worth	32.83
TX	Houston	29.98
TX	Laredo	27.53
TX	Lubbock	33.65
TX	Midland-Odessa	31.93
TX	Port Arthur	29.95
TX	San Antonio	29.53
TX	Waco	31.62
UT	Bryce Canyon	37.70
UT	Cedar City	37.70
UT	Salt Lake City	40.77
VA	Norfolk	36.90
VA	Richmond	37.50
VA	Roanoke	37.32
VT	Burlington	44.47
WA	Olympia	46.97
WA	Seattle-Tacoma	47.45
WA	Spokane	47.63
WA	Yakima	46.57
WI	Eau Claire	44.87
WI	Green Bay	44.48
WI	La Crosse	43.87
WI	Madison	43.13
WI	Milwaukee	42.95

State	City	Latitude in Degrees
WV	Charleston	38.37
WV	Huntington	38.37
WY	Casper	42.92
WY	Cheyenne	41.15
WY	Rock Springs	41.60
WY	Sheridan	44.77

Appendix B

Conversion Table For Roof Pitch

(You may choose whatever unit of distance is convenient for you
as long as it is the same horizontally and vertically)

Pitch (Rise per horizontal distance)	Angle in Degrees
1 in 24	2
2 in 24	5
3 in 24	7
4 in 24	9
5 in 24	12
6 in 24	14
7 in 24	16
8 in 24	18
9 in 24	21
10 in 24	23
11 in 24	25
12 in 24	27
13 in 24	28
14 in 24	30
15 in 24	32
16 in 24	34
17 in 24	35
18 in 24	37
19 in 24	38
20 in 24	40
21 in 24	41
22 in 24	43
23 in 24	44
24 in 24	45

Index